TWO LEAVES AND A BUD

Tales of a Ceylon Tea Planter

Written By

"Kelavan"

With Best Wishes,
Kelavan.
24th July 2001

First published by
Bloozoo Publishing, Bloozoo House, Chandlers Row, Port Lane,
Colchester, Essex, CO1 2HG
Website: bloozoo.com
E-mail: info@bloozoo.com

Second  impression
150 copies

British Library Cataloguing in Publication Data:
A catalogue record for this is available from the British Library

ISBN 1-903649-00-5

3 5 7 9 10 8 6 4 2

Printed in Great Britain by
Palladian Press, Colchester, Essex, CO1 2HG

Two Leaves and a Bud.

To all the people of Spring Valley, of every rank or grade, and to their families, who were there during the years 1938 to 1971, and to those who followed them.
Also to Rosemary and my children and grandchildren, I dedicate these pages.

CONTENTS

List of Illustrations & Maps.

Frontispiece Two Leaves & a Bud.

PREFACE

The British Tea Planter is an extinct species in Asia. He no longer exists; unless one or two very rare specimens are hiding away in some remote outpost. If this is so then their living and working conditions must be vastly different from those of the fading days of the British Empire and its immediate aftermath.

From time to time members of my family have urged me to put down on paper details of what they seem to think must have been a life of exciting times and experiences. I resisted their urgings because I felt I lacked the literary skill and, even more, feared accusations of blowing my own trumpet. furthermore there are others who could make much better job of it.

I suspect that, in fact, my planting life was very normal and mundane. There were no great adventures. It was this very normality more than anything else, which caused me to reach the decision to go ahead with "TWO LEAVES AND A BUD". Added to which nobody else seems to have taken up their pen to record this tiny part of history, which can never be repeated, about a unique way of life which has gone for ever.

Perhaps one day, before the memory has been lost for ever in the mists of time, our grandchildren, and their children will read these pages and learn a litle bit about how we Ceylon Planters and our wives lived and worked during the first half of the twentieh century,

Many generations of my family, and my wife's, lived abroad for much of their lives serving what was then the great British Empire. We continued the tradition. Our children were both born in Ceylon (now Sri Lanka), albeit after independence, and spent their childhood on a tea estate. Times have changed and they and the grandchildren must lead a different lifestyle.

Contrary to expectations I have very much enjoyed writing this litle book which I see as a sort of epitaph to a more spacious and colourful age. Fortunately a number of letters and documents have survived. These have helped immensely in refreshing my memory. I have tried to maintain my objective of writing about my life with the tea bushes, and to keep the story free from technicalities and jargon. However, in an attempt to make the tale more interesting I have digressed. The natureof the digressions will be easily apparent to the reader. The opinions expressed are entirely my own and I take full responsibility for errors and omissions concerning names and dates. Never having been a dab hand at spelling or grammar I hope that most of my mistakes were spotted and corrected by those kind people who helped check the script. The most difficult part has been wrestling with two fingered exercises on the typewriter.

Should any Sinhalese or Tamil ladies or gentlemen see these pages and be unhappy about comments I have made about their generality, may I say that the character of a people, or even of an individual, is a subject concerning which flattery is more acceptable than frankness. Also the opinions were mostly expressed when I was young and inexperienced. In this respect I beg to quote John Davy's view*:-

"This is a very general statement. It is intentionally so; I do not feel qualified to make one more particular and precise. It is difficult to know one's self; to know completely a bosom friend; how much more is it to know a people and pronounce on their character, and especially such a people as the Singalese, with whom our intercourse has been slight and our opportunities of judging extremely imperfect".

* John Davy MD., FRS., Author of "AN ACCOUNT OF THE INTERIOR OF CEYLON AND OF ITS INHABITANTS ETC" (1821)

Hansa, or Uva Goose, the mythical Goose of Uva.

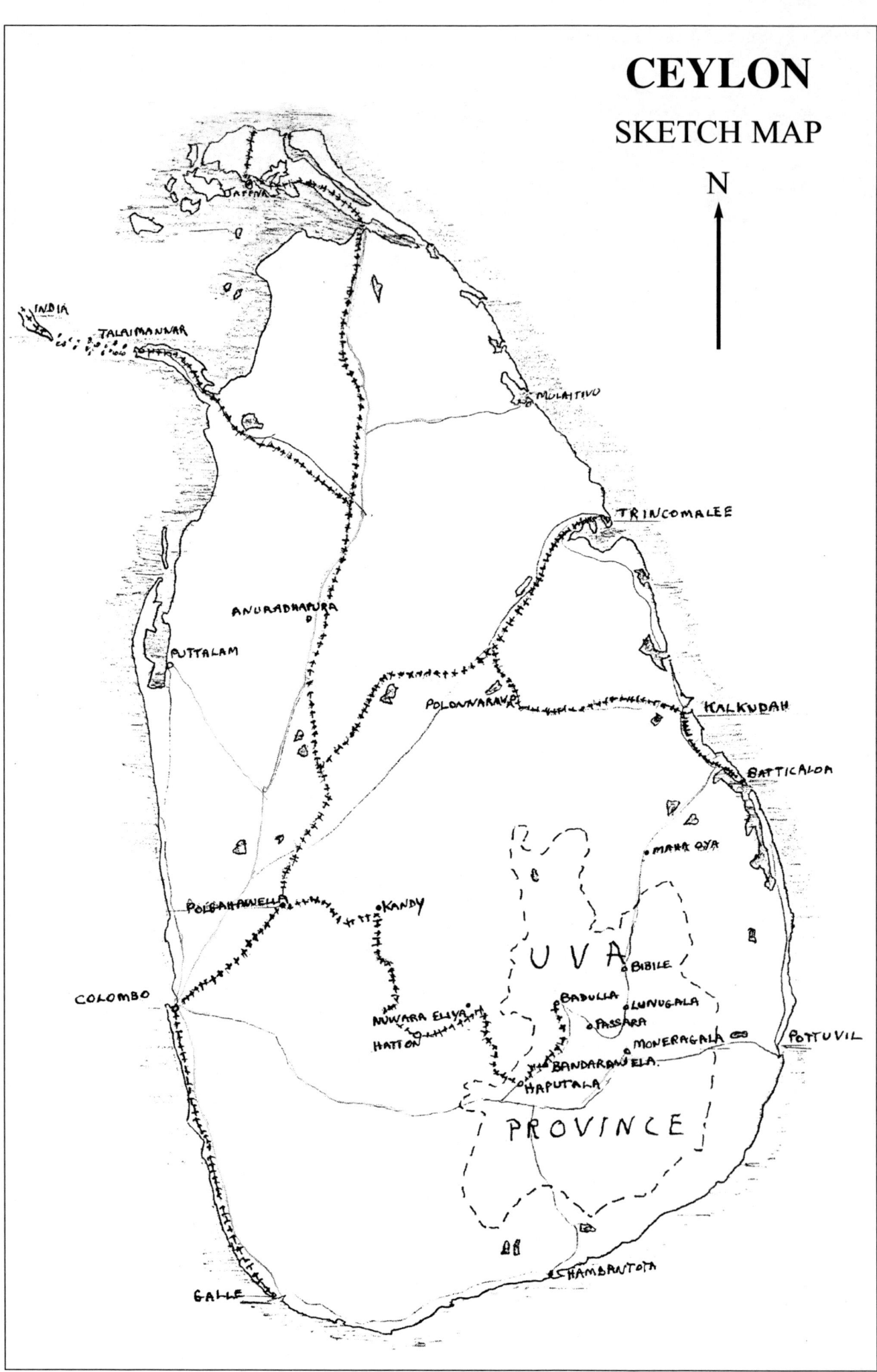

CEYLON

SKETCH MAP

N

JAFFNA

INDIA

TALAIMANNAR

MULAITIVO

TRINCOMALEE

ANURADHAPURA

PUTTALAM

POLONNARAVP

KALKUDAH

BATTICALOA

MAHA OYA

POLGAHAWELA

KANDY

U V A

BIBILE

BADULLA

LUNUGALA

NUWARA ELIYA

PASSARA

MONERAGALA

POTTUVIL

HATTON

BANDARAWELA

HAPUTALA

PROVINCE

COLOMBO

HAMBANTOTA

GALLE

CHAPTER 1

CREEPING

"Now Tea's a tricky, skilled affair, Demanding discipline and care.
The little Cuppa tree is grown
Up in the hills, a rainy zone.
The tea trees on the misty mount
Are far too numerous to count;
But each is tended, come what may,
By someone every seventh day.
It must be tended wet or dry;
You will be spared the reasons why-
But briefly, if you let it be,
The jungle grows instead of tea.
While if the soil you fail to nurse,
You get a desert which is worse..........'

A.P. HERBERT.

As a boy I was an avid reader of tales of the North West Frontier and of Himalayan exploration. Among favourite authors were Kipling, Diver, Francis Younghusband, "Ganpat", Blackwoods Magazine and others whose names are forgotten. It became my ambition to join a Gurkha regiment in the Indian Army and to serve on the North West Frontier. This was prompted,to some extent, because most of my family were obsessed by the sea. I wanted to be different. Grandfather had served time, and become a Master Mariner, in East Indiamen and Blackwall frigates; fine sailing ships trading to India and beyond. My father had spent more than half a life-time at sea; from an apprenticeship in sail with voyages to and from Australia, and round Cape Horn, to the command of passenger liners of the Orient Steam Navigation Company, plying between Britain and Australia through the Suez Canal. Three of my five brothers also went to sea after training as cadets in that famous old wooden wall H.M.S. "Conway"; now alas no more.

I was determined to break with tradition and at the same time to go far away from what I regarded as the tame life in England. However to be selected for training for a commission in the Indian Army good eyesight was essential. I had short sight. Also the cost of attending the R.M.C. Sandhurst was beyond my parents ability, for there were two sisters as well as myself and five brothers to educate and send out into the world. An elder brother was already at Sandhurst and with such a large family the fees for a second just could not be afforded. Then came ideas of joining the Indian Forestry Department or The Survey of India, but these too proved to be out of the question. Subsequently, as fortune would have it, and thanks to Adolf Hitler and Emperor Hirohito, I was destined to have the honour to serve in The 9th. Gurkha Rifles,Indian Army, on the North West Frontier.

After I left school at Easter 1935 my father arranged for me to be interviewed by Colonel Garbett, a Nabob of The Indian Tea Association who had, I believe, commanded the Sirmoor Valley Light Horse in additiion to running a number of tea gardens and companies in Assam and other Indian tea growing districts. Colonel Garbett's brother, Captain Garbett R.N. was a colleague of my father's at The Meteorological Office. Another Garbett brother was Archbishop of York. I remember attending children's parties at the bishops palace near Southwark cathedral when he was in charge of that diocese.

Anyway the day of the interview arrived. I presented myself at the offices in Mincing Lane and was ushered into the great man's presence. I was seventeen, had just left school,and was wearing my "best suit" which had been handed down through three older brothers as each outgrew it. The colonel sat me down on the opposite side of his large desk and the interview seemed to be going quite well - at least I thought so, - until suddenly there was a tinkle of coins falling through a tear in my trouser pocket and rolling across the floor. I was aghast with embarassment and hurriedly bent down to recover them. At the same time Colonel Garbett came round his desk and bent down to pick up a coin. Our heads met with an audible thump. So much for my first interview. Actually Garbett took it all very well and much to my relief laughed heartily. He told me that Planter recruits for Assam Companies were only taken on from the age of 21. All had to have practical experience in engineering, and he suggested that if I went as an apprentice to the tea machinery manufacturing works of Davidson and Company, of Belfast, he would be glad to see me again in three years time.

In those days three years sounded like a lifetime. I was far too impatient to wait so long. Fortunately for me, a Ceylon Tea Planter relation who was married to my mother's step sister, was at Home on leave,and when I heard about this I persuaded my father to take me to see him. We met at the Fowlers leave flat in Jermyn Street. Felix, or 'Flabby', as he was known amongst the planting fraternity, on account of his invariably relaxed and easy-going stance, and his wife Dolly, my aunt, were both delightful and told me all about their estate in Ceylon. About Phileda their pet muntjak deer, their dogs and other pets, and I was soon entranced with their descriptions of the country and its people. Yapame estate, in the Lungala district of the eastern situated Uva Province, actually belonged to a Colonel Isham who visited Ceylon only once every few years and left Flabby a pretty free hand in running the property. After several visits to the Jermyn Street flat it was agreed that I should go out to Ceylon in October. This would give the Fowlers a couple of months back on the estate before my arrival. Dolly had returned to England some months ahead of her husband and had spent a long time at the Hospital for Tropical Diseases where she had been treated for 'sprue'. This very unpleasant illness was fairly common then. Aunt Dolly had almost wasted away with anaemia, diarrhoea,and other unpleasant symptoms. She had the appearance of a wizened little mon-key, but was always full of amusing tales and laughter. 'Sprue' was said to be caused by droppings from termites which chewed the wooden cielings of the old bungalows. The droppings fell into food etc. This resulted in the wooden cielings being replaced by asbestos sheets. Certainly after that sprue became much less prevelant. I never heard of asbestos poisoning out there.

The arrangement was the usual one for Creepers in those days. My parents would pay the Fowlers

£100 to cover my keep for six months. During that time I would live with them in the Yapame bungalow and would be shown the ropes. A basic knowledge of the Tamil language would have to be learned from a 'munshi'; a 'babu' would educate me in the basics of estate book-keeping methods, and the field staff of 'kanakapillais' would show me how the agricultural works and the tea plucking are done. By the end of six months it was expected that a creeper should be capable of taking on a junior planter's job of running a small division, overlooked by a more senior assistant or by an estate superintendent. A junior planter is known as an 'SD', or 'sinna durai'. The equivalent of 'chota sahib', or literally 'small master'. A 'creeper' on a tea estate in Ceylon or South India is the term used for a learner and is used in the same way as 'griffin', was used in the rest of the sub continent.

Time began to fly by as my father had me kitted out by his tailors, Monnereys, of London, Liverpool and Southampton. This excellent firm supplied practically everything that was needed, from bedding to liqueur glasses. Amongst other articles of clothing they made for me was a beautiful tussore silk suit. This was eaten by termites when put into an estate store during the war. However we still have the blankets and the liqueur glasses and several pieces of crockery which have travelled the world for over 55 years.

Early in October my parents accompanied me to Tilbury and came aboard the "Oronsay" to say goodbye. my father was delighted to set foot on board an Orient Line ship again. The captain,Commadore Matheson,had served under him in the old "Otranto" before the 1914 - 18 war and promised to keep an eye on me during the voyage. "Oronsay", was torpedoed in October 1942 off the west coast of Africa when serving as a troop ship. Another fine "Oronsay" was built for the company in 1951.

I said goodbye to my dear mother with genuine sadness, but really I was delighted to be on my way at last,to what I expected would be a new and exciting life. The Orient Line had granted special concessions to the son of one of their old captains and I have always understood that the cost of my passage was £10. Britain was just beginning to recover from the disasterous world slump of the early 1930s, and many passengers who would have travelled first class in better times were now going second or third. Tourist class had not been invented. I had a second class,two berth,cabin to myself and there was always a large bowl of assorted fresh fruit in my cabin. I was surprised to find that Colonel Garbett was on board, on his way to see his company's estates in Ceylon and India. Another first class passenger who seemed to take an interest in me was W.J. Rettie. Little did I then know that in years to come I would succeed to the management of the renowned Spring Valley estate where Wilfred Rettie and his more famous father A.T.Rettie had replaced the coffee bushes with tea, and had pioneered many new agricultural practices.

There were several young planters on board , returning from their first leaves. Also several well known cricketers on their way to join the M.C.C. in Australia. I was fortunate in being able to join them in playing deck cricket. I do not remember that there were any young single girls amongst the passengers. During the first few days I was home sick rather than sea sick and there is absolutely no doubt that I was very very green so far as worldly matters, or any other matters, were concerned. My experience of life outside home and school was nil. I remember being lectured by a middle-aged Australian lady about the sinful lives led by Europeans in Asia. According to her we were all destined for a life of sloth and cohabitation with native concubines. Actually the sea voyage of about three weeks provided an opportunity for meeting and talking to old hands, and to gradually coming to terms with tropical conditions. Nowadays a youngster going east, and there are not many who go for long, is thrown straight in at the deep-end without a moment to acclimatise, and without any opportunity for initiation or advice about local customs or behaviour from experienced compatriots.

Other memories of the voyage out are the sight of the plumed and cockaded Alpini troops embarking at Naples for the Abyssinian war. The de Lessops statue at the entrance to the Suez Canal (now no more),and the jeers and cat-calls of British troops along the banks of the canal as the 'Iteye' troopships passed through. British ships which were outward bound, such as ours, were greeted with shouts of 'You're going the wrong way', and 'Get your knees brown'. The barren rocks of Aden seemed just that, but were followed by a delightfully cool Indian Ocean during the North East Monsoon, and a glimpse from the deck of the lovely palm covered island of Minicoy the day before arriving at Colombo.

The 'chit' system was quite new to me. To be able to get a drink at the bar by merely signing a piece of paper seemed too good to be true. I was able to slake my thirst, and the thirsts of my new friends, in the heat of the Red Sea and Indian Ocean with lashings of cool beer, without regard to cost. The fact that the drinks were duty free and so very cheap did not prevent me getting a shock when my bar bill was presented the evening before arrival. I discovered that my liabilities on board considerably exceeded my liquid assets. Other assets I had none. Somehow a kindly member of the Burma Civil Service noted my predicament and insisted on lending me £10 to cover my expenses. This was eventually accepted with embarrassment as being the only solution. £10 was a considerable sum in those days and I had every intention of repaying Mr. Watson (I have never forgotten his name), immediately after receiving my allowance on arrival at Yapame. Unfortunately I mislaid his Rangoon address. About eight years later, in an hotel in Delhi during the war, I ran into Watson and recognised him immediately. When I introduced myself and offered to repay the debt with interest Watson denied all knowledge of ever having met me, let alone having given the loan!

Soon after the ship moored in Colombo harbour a Ceylonese came aboard and handed me a letter from Flabby Fowler. Owing to various emergencies up-country he was unable to come and meet me. However he enclosed some money for the train journey instructions for sending the baggage up to the estate in the estate lorry which would be found near the customs, money for some shopping to be done in The Fort for my Aunt, and advice to go to the G.O.H., (Grand Oriental Hotel), near the jetty, where I could book my train ticket and a sleeping berth on the night mail due to leave that evening for Badulla, at the end of the line, where he would meet me the following morning.

I went ashore in a harbour launch, passed through customs, found the lorry marked, Yapame Estate, loaded my luggage on it, and crossed the street to the G.O.H. where I joined some of my ship-board friends in the Palm Court. I was agog at the strange surroundings and the humid heat. I noted down my first impressions of "raucous thieving crows, shady tree-lined streets, fine shops, tramcars, rickshaws and hooting taxis of the open variety(now only seen on American films of the 1920,s), and people of many races. Most of the men are dressed in long white skirt like costumes, their hair tied in buns. on top of the head like victorian ladies, but surmounted by half-moon shaped tortoiseshell combs. In the G.O.H. where we sat under the whirring cieling fans, it was much cooler. This is the Fort or business area of the city. It was built on the site of the old Potuguese and Dutch Fortresses. Not far away is the Pettah, the native quarter".

We slaked our thirsts with ice cold Nuwara Eliya beer (pronounced Newralia), or long glasses of fresh lime juice with ice cubes floating on top. The first meal in Ceylon was taken in the hotel's "pop" restaurant"; delicious crayfish in mayonnaise followed by a delectable Sinhalese curry. Afterwards a visit was paid to Whiteaway Laidlaw's excellent emporium just round the corner in York Street. This establishment was closed in the 1960s, alas. The beautiful ingasaman trees which heavily shaded the Fort's streets, the rickshaws and the buns and tortoiseshell combs have all disappeared with the march of time. Nearly all the male, and not a few of the female city inhabitants now wear trousers. Whiteaway Laidlaw's was locally known as 'Right Away and Paid For' because it was about the only place which insisted on cash payments. On referring to old letters to my parents I see that it was the kind and long suffering Mr.Watson who guided me to the right shops, and advised me on the right sort of shorts and khaki pith sun helmet to buy. He took me on to Cave's bookshop where I bought a simple guide to the Tamil language called "Udday Uppar". Literally translated this means "oh! My father", and is roughly the equivalent of "Good Heavens", an expression of surprise. Later I acquired another little book, equally useful if not entirely grammatical, titled "Learn Yourself Sinhalese".

Soon it was time to catch the Night Mail and a rickshaw took me with my suitcase through the dark, noisy and crowded streets,to the Fort Station. A porter took my bag, hoisted it onto his head and ran ahead of me, threading his way through the crowds, and past many sleeping bodies on the platform, to a first class sleeping compartment. According to a letter to my parents dated 6th November 1935,: "I boarded the train at 9.40 pm on Saturday. A Mr.Armitage who was in the bottom sleeping berth, was going to Hatton, and knows Uncle Felix quite well. He had also sailed in an Orient ship under your command Dad. It was very hot and noisy but I managed to sleep and woke up at 4 am absoluteley frozen. Mr.Armitage was dressing and got out soon after. It was light just after Nanu-Oya, and I saw huge mountains on either side with cascades of water rushing down them, and the sides covered in tea bushes. The train kept up and up until it reached the summit. I think that is over 6,400 ft above the sea. Then it went steadily down hill and I had a marvellous view to the east with all the valleys covered in morning mist. I got some breakfast in the restaurant car and we arrived at Badulla at 11.30 .

The journey from Colombo to Badulla of 181 miles had taken nearly 14 hours. Actually the train had arrived two hours late and the scheduled time for the journey, to this day, is almost 12 hours. There are numerous wayside halts for the loading and unloading of mail and passengers, and the terrain prohibits speedy travel between stations. The summit is reached after passing through the highland planting districts of Dickoya and Dimbula, with their luscious damp green tea covering the ridges and valleys which, in turn, are often capped by thick jungle or white clouds. The engineering of this line is said to equal the mountain railways of Norway and Switzerland and the final section to Badulla was only completed in 1926. Between the numerous tunnels bored through the mountain sides, the rails bridge many chasms with girders placed from rock to rock and the lines laid over them, whilst torrents surge through the gaps below. The winding is such that, in places, to accomplish a distance of one mile direct, the train traverses six miles of railway. This is done in a fashion so circuitous that a straight line would cross the rails no less than nine times.

From the summit the descent to Badulla, of about 4,000.feet is gradually accomplished with many more twists and turns, plunging into tunnels and over rushing streams. At Idalgashinna, where I was joined in my compartment by a young planter whose mode of arrival was surprising, of which more anon, the line runs along a ridge giving stupendous views on either side. The scenery in Ceylon frequently causes euphoric comment and James Taylor,the pioneer who produced the first Ceylon tea to be marketed (when the island's great coffee industry was devastated by the fungus disease 'hemeleia vestatrix'), was among the admirers when he wrote Home in 1858; "You will think I write a lot about the scenery but if you saw it you would not think I said too much".

That morning was typical of the start of the North East Monsoon. The air was absolutely clear, fresh and bracing. The sort of air that makes you want to dash out and savour the countryside. Heavy clouds would build up behind the hills by noon, and later descend to shut out the view with thick mist, as thick as a London fog, through which a steady rainfall would be likely to develop and continue late into the evening or night. At that moment we looked down from a great height towards the South and East. In fact the southern quarter of the island from Yala to Galle was laid out far below like a great green and blue map. There was the lowland country thickly covered by jungle, diversified by rocky hills rising abruptly from the flat land, and interspersed by an occasional wandering silver line indicating a river. Away on the horizon, some sixty miles as the crow flies, was the sea, visible as a broad band of silver illuminated by the rising sun. On the other side lay the rolling plains of Uva with their grass or "patna" covered slopes over and under the green ladders of terraced paddy fields set against the open hillsides. Thatched houses of the villagers showed through the wooded valleys. According to Robert Knox in his "An Historical Relation of The Island of Ceylon", first published in 1681; "The Province of Ouvah is a country well watered, the Land not smooth, neither the hills very high, wood very scarce but what they plant about their homes...... Rice is more plenty here than most other things". When the train halted, which was frequently, the best remembered sound was of the villagers high pitched "hoo" cries urging their buffaloes in the paddy fields.

I was met at Badulla by my uncle and taken to the bungalow on Yapame an hours drive on narrow mountain roads through tea and rubber estates, through the crowded narrow bazaar of Passara town, and past patna covered hills with occasional views of the lowcountry jungle and glimpses of the distant sea.

Owing to the dearth of junior planting billets my "creeping" lasted over one and a half years rather than the usual six months. In the end this was probably all to the good as I was able to pick up a lot of knowledge about work in the factory and office not usually available to creepers or junior SDs. In fact the Head Teamaker, in charge of the factory, was sent on leave to India and I was told to take charge in his absence for a couple of months. This way I picked up quite a "few tricks of the trade". In the same way the Chief Clerk, Mr. Dawson, was sent off to his home in Jaffna for a long holiday and I had to act for him. I made an awful mess of the books and spent many nights burning the midnight oil trying to enter up journal and ledger and to balance the monthly accounts.

My uncle must have been very long suffering, and probably despaired of ever getting me fixed up in an appointment as an SD. However I certainly did not relish the thought of having to return Home, with the assumption of failure. Many estates were then running at a loss and not a few planters were "on the cart road" (out of a billet). Some left the island and others took jobs as Teamakers and the like usually reserved for Ceylonese of the Estate Staff. This was a period of world over production of tea. To overcome this an International Tea Regulating Board had been set up and the tea producing countries were given quotas based on a proportion of production during preceding years. Within each country the local boards then allocated their own quotas to the various companies and estates. These quotas were a cause of much heated debate and discussion, many planters considering they were unfairly treated and should have had a better allocation. Whether this was really so or not all estates had to cut production, which meant employment of fewer staff and labour.

Within a week of my arrival I had an unexpected trip to the East Coast. Uncle Flabby was a Justice of the Peace, and a telegram came saying that a rogue elephant was causing havoc near Chenkaladi, a village in the Eastern Province, and had killed or wounded some road workers of the Public Works Department (PWD), Presumably the officials in that province were all otherwise engaged so they sent for Flabby. We set off at dawn with rifles and shotguns. After descending the jungle lined pass to Bibile we travelled through miles of low country jungle in the little Flying Standard car. We were stopped by a crowd near the Irrigation Department Lines at the tiny hamlet of Karadiyanaru. The only dwellings then consisted of the lines and the overseers bungalow. The name of this Place could be translated as "The river where the bears live".

The elephant had killed one of the road gang, had pulled thatch off one of the roofs and had torn up a 'kadjan', fence. It had wandered about near the scene of its attack for a time, but when we arrived it was nowhere to be seen. My uncle did whatever his duties as a J.P. required and had the squashed remains of the dead man removed from the road with a "mamoty', (inverted local spade), and put in a sack for disposal. After this unpleasant duty, and talking to the people, who were a poor lot suffering from repeated malarial attacks like all the people of that area before the discovery of modern preventative treatment, we motored on to the lovely coconut palm fringed bay at Kalkudah. Here we had a refreshing and cleansing swim before returning home along the jungle road by night. We were thankfull there were no wild elephants on the road. Driving through the jungle at night was fascinating. Sometimes we did come across wild elephants and always mysterious eyes of some wild beast or bird reflected back from the headlights. On one occasion we topped a rise to see a huge d.ark object just down the road. We halted the car and waited with the engine quietly ticking over - ready for a quick turn round, or a dash past the object if opportunity permitted, as we were convinced it was an elephant - After waiting for a long time it still made no move, so we let in the clutch and advanced cautiously and as we drew near we saw that the "wild elephant" was a large steam roller and tar boiler left by the roadside. The eyes of night jars would shine from the road until we were almost upon them when they would take flight just in time to miss the car's windscreen. This often gave us a fright.

After being at Yapame for several weeks I wrote home to my parents saying :- "Every morning I am woken at 5.30 and go down to the factory for muster at 6. The coolies line up in a semi-circle, two deep, and are sent off in gangs under charge of a "kangany' (foreman), to various tasks; pruning, forking, terracing, weeding, manuring, plucking etc. I enter the number of people sent to each job on a printed Muster List, and put down the total pounds of green tea harvested the previous day, the average pounds brought in by each plucker details of other works completed, and the rainfall. Last night there were 4.60 inches of rain. I come back to the bungalow for early tea at about 6.30 or so and go out again, to the fields, soon after 7. The leaf plucked is weighed at 9am, noon and 4.30pm. Every time the leaf is weighed each plucker has her "tundu" marked with the number of pounds plucked. Each field needs to have the crop plucked once a week throughout the year, and if a plucker brings coarse leaf this has to be picked out and thrown away before her basket is weighed. Good pluckers have very nimble fingers which seem to move like lightning. They take "two leaves and a bud" and snap off any coarse leaves which are above the level table of the bush, and discard them. They also discard leaves called "banjies" which will not grow new shoots. Side branches must not be taken below 'the level,' so that the spread of the bushes will increase. I come in to breakfast after the noon weighing, then go out again at about 2pm. After coming back for tea at 4 (though not always), I go to enter the "names" in the pocket check-roll. After that I have to balance the check roll and enter the costs of the various works done during the day. This takes an awfully long time and I should think my maths has improved well over 50% since I came here, because when I was learning how to keep it Dawson, the jaffna Tamil clerk looked over my shoulder and I daren't make a mistake in front of him.

"Today Uncle Felix has been showing me the big check roll which looks a very complicated affair. One thing is that you can easily tell when you have made a mistake, because every page should balance itself. If it doesn't then it is a case of finding where the error is on that page. I will have to calculate and enter all the wages in this book from next month. I have just finished doing the rice account. Each worker gets so many "soondus" (measures) of rice per week, according to their position, age and sex etc; for example, a kangany gets seven measures whether he has worked six days in the week or not. A coolie gets six measures if he has worked six days, and one measure less for each. day he fails to turn up. A woman gets five measures with one off for each day absent, and a boy gets 4 measures. There are 32 measures in a bushel of dry rice. I shall issue the rice tomorrow.

Kalkudah Bay from the rest house, 1935.

Out on a snipe shoot, 1936.

East Coast villagers, 1935.

East Coast Ferry, 1936.

There is a marvellous view from my bedroom window,of high mountains and fast streams with waterfalls rushing down. You can't imagine what huge meals we eat when we come in. Uncle Felix is very energetic. Most of his estate is about 2,000 feet above the bungalow and he often goes up on his horse twice in a day. He is also in charge of two rubber estates, El Dorado and Cocoawatte which he visits periodically and has his horse sent ahead to meet him there. This bungalow is 2,700 ft. above sea level and the factory is 600 ft. below. We are within the malaria belt but owing to very strict and regular oiling of all stagnant water there is very little of this illness on the estate. The factory machinery is driven by water power from a pelton turbine which is supplied from a dam in the garden. This is kept filled by a watercourse which comes from a big stream over a mile away up the estate. The electric lights are also powered by this means. In order to conserve water in dry weather, so that, the dam can refill, the pelton is shut down when we go to bed.

I am beginning to pick up a few short Tamil sentences but they speak so fast it is difficult to follow. The Kannakapoolie (Head Overseer) is very helpful and explains most things. The pruning kangany, Arunasalem, is showing me how to prune. You have to have a very sharp knife and be able to prune 250 bushes properly a day. He seems a good chap and the pruners all do as he tells. He gives his instructions quietly. Uncle Felix says he is high caste and that helps his influence. I work in Keenagoda division which has nearly 200 acres of tea and some eucalyptus gums, and about 7 acres of rubber which is not being tapped now. The factory is in this division and the rice stores, carpenters and blacksmiths shops, and the stables for the bullocks which pull the carts to collect the tea from around the estate. The leaf from Yapame Division is sent down to the end of our road by wire cable shoot where the bullock carts collect it and take it to the factory.

Last night a young chap named Walter Ogilvy called. He is an SD on Swinton Estate. He says this is an outlying division of Hopton, which is the biggest estate in Lunugala District. There are two other SDs there, Macara and Drury, but they live a long way off at the other end of Hopton. Ogilvy has been out here about five years and seems a very nice fellow. He has a small dog which travels round with him sitting on the tank of his motor cycle. He is getting his first Home Leave next spring. There are no other young fellows but he has asked me to come to the club to play tennis on Wednesday".

The small planting district of Lungala is situated in a long valley below the high Madulsima Range which is joined to the higher Namunukula Range, by a low saddle on which the small town of Passara is situated. Lunugala contains some of the most easterly tea estates. Beyond it the country falls away down the Bibile Pass to the low-country which in those days was nearly all jungle, except for a narrow strip of coconut plantations along the eastern seaboard. Lunugala was served by a small tennis club. This little club was the only place where planters and others could meet on common ground, Membership at that time consisted of nine planters only. Five of these were estate superintendents known as PDs or "Periya Durais", ((the equivelent of "bura sahibs"), and four SDs (Sinna Durais). the other eligible people in the district were the Sinhalese doctor in charge of the goverment run hospital, and the burgher doctor in charge of the hospital on Hopton estate, but I do not think they were members, at any rate I never saw them there.

Club day was Wednesday. The two tennis courts were then prepared for play, the shutters of the small corrugated iron roofed verandah were lowered, tables and chairs set out, and the small bar opened up. By about half past three in the afternoon the club's only servant or "boy" would have everything ready. He went by the name of Mariasoosay and lived in a small house alongside the little club. There would also be four or five small "podians" (small boys), sent from Yapame to act as ball boys. These would be dressed in khaki shorts and shirts topped by red scarves tied round their heads turban fashion.

At Yapame Estate - 1936

Yapame bungalow.

The garden and dam.

After a swim.
Left to right: Walter Ogilvy, Cynthia Massey and T. G. Macara.

Members supplied tea by rotation. The "duraisarnais" (ladies), usually vieing with one another to provide the richest and most elaborate cakes. This had something of a dual effect as hungry young bachelors, such as myself, took the opportunity for a really good tuck in, but when it came to our turn to provide tea the SDs found the cost to be prohibitive, or else their menages incapable of baking cakes. The sun set early throughout the year between 6 and 6.30, and darkness followed with the usual tropical suddenness. The PDs usually departed soon after, preferring to take their sundowners in the comfort of their bungalows. Most of the small band of SDs stayed on and the whisky bottle would be placed on the bar for us all to help ourselves. At the end of the evening we would all sign a chit for the total amount consumed and all were charged equally. It does not require mathematical expertise to reckon that the heavier drinkers had an advantage, and although the cost of Whisky then was six rupees a bottle (say 9 old English shillings or 45 pence of our modern currency), SDs salaries were far from being princely.

My first visits to the club were made with much trepidation. It was an easy walk from Yapapme as the club was situated on the boundary of my Kinagodde division with Swinton. The young SDs, and specially Walter Ogilvy, made me very welcome though I was the youngest by five or six years. No doubt any new face made a welcome change. I was in considerable awe of the PDs, especially of such characters as "have a dram" Ruxton, the large and florid manager of Hopton who hailed from Aberdeenshire, and of Carson-Parker of Shawlands estate. My impression was that with one exception they were not on polite speaking terms. The exception was dear old Saibo Sim of Kehelwatte estate, who would go from one to another chatting to each with a humorous twinkle in his eyes. The Fowlers seldom attended, and the Fanshawes, who owned Park Estate, generally kept aloof but sent their "appu" (cook) with the tea when their turn came.

"Saibo" Sim was a frequent visitor to Yapame and on fine dry nights we would all sit outside under the stars for our drinks before dinner and I would be enthralled by the yarns about the old days before the 1914 to 1918 war, and about the characters of the old coffee planters. In his younger days Saibo had cycled from Badulla, the capital of Uva Province, to Baticalloa, the capital of Eastern Province, a journey of about 120 miles most of which was through thick jungle. This was no mean feat in those days before the road was improved for motors. The usual mode of travel then was by bullock coach, taking about three days. There were small, very primitive, resthouses in the jungle spaced about fifteen miles apart. Saibo had several close shaves with wild animals, and drank all the beer in the resthouses on the way down, which he much regretted on the very hot. uphill return journey. My uncle had many reminiscences of the South African War, and stories about such adventures as sailing in "katamarans" off the Ceylon coast. The speed of these boats in a fair wind was such that on one occasion off the southern extremity of the island, at Point de Galle, they overhauled a P & O liner, with two members of the katamarans crew standing out on the windward outrigger to keep it from lifting out of the sea. On another occasion a monster octopus nearly pulled their boat down with its immensely long tentacles. Fortunately they had a large "vetta catty" (bill-hook like knife). and were able to cut themselves adrift. I regret not having noted many other yarns about opening up and planting the early rubber estates, chasing and destroying rogue elephants, and campaigning with the Ceylon Mounted Rifles in the Boer War. Elephants were declared "rogue" by government order when they had caused much damage to humans, habitation and villagers crops. They were often old animals which had left the herd, and perhaps had been wounded by village watchmen defending their precious crops. As a result they turned vicious and dangerous and had to be destroyed

Dollie Fowler's maiden name was Bean. She had an older and much loved sister Minnie, who married Carl Drieisma in 1897. Minnie and Dollie had gone to South Africa with a ballet dancing troupe, and it is thought Minnie met Carl there. Shortly after the end of the South African War in

1902 Carl was in Ceylon and was involved with the opening of several rubber estates in the Kelani Valley and Kalutara Districts in the South West of the island. Amongst these was the Frocester Rubber Estate. He also acquired interests in several tea properties including the Ceylon Provincial Estates and the Doomoo Tea Company in Madulsima. Carl and Minnie moved to Glassaugh, one of the Ceylon Provincial company's properties, at Nanu Oya not far from Nuwara Eliya, in about 1908 or 1909, presumably to be in the highlands for Minnies health. There is a charming photograph of Minnie riding her pony in the tea. Most of Carl's money was invested in the rubber estates which were being rapidly expanded. Many were giving first rate profits at that time, rubber only recently having been established as a plantation crop in South Asia owing to the inability of the crop from the wild rubber trees in South America to meet the demands of the market. Carl did so well that in about 1912 he bought a large estate and mansion in Devonshire and retired there with Minnie before the first world war. Earlier, perhaps during the Boer war, Felix Fowler had met Dollie at The Cape. Dollie was about to go to join the Drielsmas in Ceylon. Felix followed her there and managed to get an appointment on one of the estates through the good offices of Carl. Felix and Dollie soon married and settled down to estate life. Minnie and Dollies mother died when they were small children. Their father, the Revd. A.H.S.Bean, of Saaby, Yorkshire, re-married. My mother was the eldest of his eight children by the second marriage. Hence my reference to the Fowlers as "aunt" and 'uncle". The Fowlers and Drielsmas had no children but both were extraordinarily devoted couples. Dollie always referred to Felix as her 'Polar Bear' on account of his snow white hair. He was a very big man, a fine horseman and a keen shot. She was tiny, when I knew her after being an invalid, from sprue and other tropical complaints which seemed to wither her away. However she was always full of fun and laughter and had plenty of guts. I have a great deal to thank them for, especially for Felix's example in management of labour, giving me a really good grounding in the job, and for his invariably straight dealings and support when in difficulty.

Every planter, or for that matter any Europeanwho works or deals with Asian labour, is well advised to learn how to behave towards them and to respect their customs and culture, which are often very strange to us, as are ours to them. I soon learned not to ask a woman for the name of her husband. It is not polite to mention the word, "husband" to a man's wife, nor for the wife to speak her husband's name. If you want to know ask someone else. I also learned not to praise a child, or a new baby, to its parents, and that it is better to say "what an ugly little monkey" or words to that effect. To notice good qualities of any sort will only arouse the jealous anger of the gods who may put a blemish on the child or bring it bad luck. Another lesson learned was to avoid, or to take no notice of workers when they ate their rice. Mealtime is sacred. To allow your- shadow to fall on their food may well cause annoyance, or pollute it in the case of high caste workers. There are many other tabus knowledge and practice of which helps towards happy relationships.

The Yapame labour force was generally contented and well behaved. Pay day which occurred monthly was a special occasion. Work was usually knocked off early to give everyone a chance to tidy up and come to the pay table in clean clothes. It was important that every worker, so far as possible, had his or her wages paid into their own hands. This way we got to know all the workers names, as well as their relationships with each other, and to recognise their faces. It also showed that the worker actually existed and was not just a name in the check-roll for someone else to get the pay. A note could be taken of a worker's health and enquiries made about a sick relative whose presence would be excused. Some planters boasted that they got through pay in record time but this was at the expense of good labour relations.

In the garden at Yapame, Felix Fowler, Dollie Fowler and Hilda Bean, 1936.

Minnie Drielsma riding in the Tea.

The PD or SD in charge of pay would be seated at a table on which the money would be arranged by denomination , and with the clerk to one side reading out each man or woman's name and the amount of balance pay due; after deduction for foodstuffs and any other items issued on credit. The kangany of each labour gang would take his turn near the table,being responsible for seeing that the members of his gang were present. on hearing their name called the man or woman would answer "aiyah" (sir) or durai (your honour) etc, and come forward with right hand outstretched, palm upward, often supported by the left hand at the wrist or underneath the right, take his or her wages, bow or salaam and depart.

At a distance from the pay table some of the workers, and perhaps several Sinhalese villagers, would have set up temporary stalls of sweetmeats, colourful.. drinks, glass bangles and the like,to catch customers whilst flush with money. Further away, and usually out of sight (because they knew if they were seen by the durai they would be sent away), a few money lenders might be waiting to collect debts, or at least some of the exorbitant interest which they charged. Another reason for insisting on individual payments was that should a man be given his wife's or his children's wages he might go straight off to the village to spend it on "toddy" or some illicit alchohol.

The transport of wages to estates for payment to hundreds or,in the case of large properties —, thousands of workers, needed considerable organisation. After completing the wages checkrolls, an analysis would be made of the actual number and denominations of notes and coins needed, so that each individual worker could be correctly paid. These details would then be sent to the bank (which might be anything from ten to 100 miles away), with instructions to have the cash ready in sealed bags, and a date and time given for it to be ready. In the peaceful days prior to the second world war, the manager of the Bank of Uva,at the Uva provincial capital of Badulla, a Mr. Queckett who incidentally was almost as deaf as a post, would travel to Colombo by train, collect all the money bags for thirty or more estates, load them into a taxi, drive to the railway station, have the money stowed under his sleeping berth on the night mail, and sleep the night away peacefully whilst lying above hundreds of thousands of rupees.

When I first went to Spring Valley in the 1930s, where the work force was around 2,500 souls, the Chief Clerk, a fine Sinhalese gentleman by the name of Philip Perera, was driven to the Uva Bank in the manager's car where he collected seven sealed bags, one for each of the estate's seven divisions. On arrival at the estate Head Office these bags would be handed to seven waiting "tappal" (mail) runners who would disperse with their loads on their heads to the divisional offices. This involved journeys of up to three miles through steep and wooded, often boulder strewn, country. It never entered anyone's head that there might be a robbery. Actually a couple of armed robberies did occur in the low-country rubber area not far from Colombo,and in one of these a planter was killed. However that was considered a different world to ours. It was not until 1947 that the system was changed on Spring Valley when a bag of money sent to the May Mallay (Upper) division was found to have been tampered with.

From the early nineteen fifties, following several brutal armed robberies in our district, we changed the system so that never less than three vehicles travelled in convoy to collect pay. Each vehicle contained a planter with a sawn-off shot gun. The time and date for wage collection was kept secret until the last moment, and the money was distributed amongst the vehicles. Routes were varied as much as possible and all the wages had to be paid out on the same day that they were collected. This fact had to be reported by telephone as soon as possible and be confirmed in writing within twenty four hours. Times were beginning to change for the worse. All this was to come about in the years ahead. Meantime we must go back to Yapame and the business of 'creeping'.

There did not seem to be many labour problems on Yapame, though doubtless unknown to me there would have been the usual family and factional quarrels. Generally this estate was working on the old patriarchal system which was widespread up to at least the 1914 - 18 Great War. Under this system the Head Kangany ensured that labour was plentiful, and usually he was the PDs right hand man. He dealt with labour problems, loaned money to the workers to help their land problems in India, and to visit relatives there, and saw the work was well done. He recruited workers from among his relatives and friends in his own and adjoining villages in South India. From the point of view of community life on the estate this had a good deal to be said for it. The Head Kangany stood by the young and inexperienced "durai" if the latter was anything of a decent fellow. But as government regulations increased, and Trades Unions were founded, the Tamil worker became, at least mentally, a very much freer man than ever before. He was getting education and, by degrees, a greater independence of thought and action. He was throwing off the patriarcal idea and mentally growing up. As could be expected he began to display some common symptoms of the young and growing - a greater independence of character-and, frequently, an exaggerated idea of his own importance. This was an excellent thing as indicating progress as a member of the human race, but was bound to have an appreciable effect on relations between employer and employee, between "cooly" and "durai". There was less blind submission from one, and more knowledge consideration and tact called for from the other. The 'durai' too changed. Motor cars and motor bicycles became generally owned, and thus the week-end habit grew. Durai spent much of his off-duty time away from the estate; off to his club and to his neighbour. This tended to make the modern durai to be in less close contact with his labour than in days gone by. So with changes on both sides labour needed very careful han- dling. A better knowledge of the language and caste and custom became necessary.

One day, not long after arriving at Yapame, I was walking through a pruned tea field near the 'cart road'; the main road from Badulla to Lunugala and beyond, when I noticed a cloud of blue bottles rise buzzing from a drain close to where I was standing, looking to see if the area had been properly weeded. On going closer there was a strong and unpleasant odour, and then I saw the body of a woman with a large, blood covered gash on the side of her head. I hurried to the office and reported to my uncle. I don't remember the rest of the story, or anything about the cause of death, but it was the first dead body I had seen and not a very nice one either. Of course it had to be dealt with speedily as in that climate burial or cremation has, for obvious reasons, to be undertaken with- in 24 hours. Most of our workers bodies were disposed of by burial and every estate has its own burial grounds.

On another occasion, when a turning place for the bullock carts was being quarried on one of the estate roads, the workers encountered a large boulder the size of a small car. They cut away the soil all round it, so that by the time of the noon meal break the rock was protruding well out from the bank. I had the area roped off and went back to the bungalow for my own meal, after warning everybody to keep clear until my return. Only about twenty minutes later there was a great noise of wailing and lamentation. We rushed to the spot to find that two of the workers had ignored all warnings and had decided to eat their meal under the shade of the rock. This had suddenly fallen upon them leaving only their extremities protruding from beneath it. I was rapidly becoming acquainted with violent death.

In those days, the mid 1930s, everyone on the estates waited eagerly for the arrival of the Home Mail. This always reached Colombo, without fail, once a week at dawn. The P & 0 steamers arrived on alternate Sundays, and the Orient Line ships came in on the Saturdays in between. Country mail reached us by noon the following day, so that when it was the turn of the Orient Line we got our mail before lunch on Sunday, and had plenty of time to read and enjoy it. Home mail was so important because we felt very cut off from Home. Nobody could dash back to Britain, or to Australia, or wherever as they can nowadays. The Imperial Airways and K.L.M. were just pioneering passenger air routes to the East. At that time those fine bi-planes named "Hanibal" and "Hasdrabel", and the like; took seven days to reach Bombay from Croydon. Each night was spent on the ground in a luxury hotel. In any case the cost of such travel was way beyond the means of all except the very rich. Nor was it considered very safe.

Once a planter got an appointment - and I still had not even a job in view -, provided he was employed by a good company or Agency House, a "Passage Money Fund', would be set up in his name, and funds would be set aside annually to accumulate sufficient, over five years, to purchase a return sea passage. A planter's first tour abroad usually lasted five years, and after that gradually reduced until, as a senior man, he might get down to tours of three years. Not many companies provided passages for wives and children. These had to be saved from a planters earnings in the majority of cases,and very often children who were sent Home to school at the age of about ten would not see their parents again, or at least their father, until they were aged fifteen. I think most of us accepted this as being perfectly normal. In fact there was really no other way. Owing to the war intervening, I did n ot get Home for ten years. My first short leave was from the Indian Army in 1945. From the late 1950s there was a major improvement in the terms for family passages. This coincided with the advent of regular 'economy' tickets on B.O.A.C. The company I worked for was a leader in agreeing to these improved conditions of service.

The Colombo wireless station broadcast programmes in English, Sinhalese and Tamil on Long Wave. These were often subject to climatic interference, specially if you were listening whilst among the mountains Up-Country. Short Wave transmission was in its infancy and reception was very unpredictable. Saibo Sim had recently brought back, from Home Leave, one of the very latest wireless sets which could be tuned to a variety of wave bands by changing coils to suit the band of your choice. I remember getting reports of the illness of King George V, in 1936, direct from Daventry, using the 16 meter band coil. Reception was poor with much crackling, fading, and atmospheric interference. We listened with our ears close to the loudspeaker. Saibo had also brought out a brand new Austin Seven tourer, the first of these tough little cars to be seen in the district.

In remote Provinces, such as Uva, planters often lived very isolated lives which could lead to great loneliness. This was specially the case with some junior SDs who could not afford, or else were seldom allowed, leave to travel about or to participate in social events. Probably it was the younger wives who found time hanging most heavily. For this reason marriage was not encouraged among young planters. Some were unable to stand the loneliness but these were usually weeded out before much damage was done.

In addition to the charge of Yapame Estate and an appointment as Visiting Agent for other estates in the Agency of Mackwoods Ltd., Flabby Fowler had charge of two nearby rubber estates, El Dorado and Cocoawatte. 'Nearby' is a relative term. El Dorado situated above the Bibile Pass, was opened up during the rubber boom of the 1920s. It was financed by Flabby, Saibo and another whose name I forget. The young trees were just coming into bearing when rubber prices crashed to rock bottom. The owners never recovered their losses and the estate was eventually sold for a song after the second world war. So much for ambitions for El Dorado.
Cocoawatte was situated in a deep valley about fifteen miles from Yapame, as the crow flies, but getting there involved a drive of some 20 miles on difficult roads, to a grassy plateau where the car had to be abandoned. From this point it was necessary to ride or walk down a steep escarpment into a hot and steamy valley. I occasionally accompanied my uncle on his visits. When we got down to

the level of the rubber there were always signs of wild elephant. This was in the form of droppings and damage such as broken branches and uprooted fencing. Walking through this rubber with nothing but a stick for protection made me feel exposed and helpless. At that time there was a Burgher Conductor in charge. Previously this had been a European SD's post, but the last three men appointed had all failed to stand the isolation. One had given up and resigned. Another had "gone native", refused to do any work and lived the life of an indolent Sinhalese peasant with the local ladies of his choice. The third had gone mad and had to be shipped Home. One of the three, I forget which, had passed his time at training himself to become an expert revolver shot. It was said that he would expect any visitor to throw playing cards into the air and he would, without fail, put a bullet through any part of the card named. These two estates have disappeared - perhaps covered in jungle - They are no longer mentioned in the Directories of Estates.

Because of the dearth of cultural facilities other than field games and outdoor sports such as shooting and fishing, creepers were usually selected for their abilities at such activities, rather than for educational qualifications. Very often a creeper would be a relation, or someone well known to a planter. This ensured that there were not many duds, as no-one was likely to reccommend anyone of doubtful character to a friend, for fear of getting the blame and having to take the consequences which could involve the cost of a passage Home. There were many more failures later on, after the war, when young men were chosen at Home by Selection Boards set up by the London based companies. Some very strange appointments were made. In retrospect I am sure that I was most fortunate in failing to get a billet after the usual six months creeping. I would have found the loneliness of an isolated bungalow unbearable at so early a stage. As it was I had a year and a half living with relations, followed by ten months in the house of my first employer, whose wife, child, and European nurse, were all with me under the same roof. This continued until they all went on Home Leave. I was then left in charge of the estate for six months. I had the comforts of their well equipped bungalow, their servants and dogs and horses to keep me company, and to give me a sense of responsibility. A senior planter was available at the end of the telephone, for advice, and visited me fortnightly to see how things were on the spot. In fact I was kept so busy that there was no time for loneliness or to notice the lack of entertainment.

Whilst creeping at Yapame there were the small Wednesday afternoon tennis gatherings at the Lunugala Club, and the occasional calls by Saibo Sim and a few other friends of the Fowlers at their bungalow. Another occasional visitor was Portland Henty, the Manager of Passara Estate situated in the district of the same name, to the South of Lunugala. Portland used to arrive on his motor cycle with his old mother riding postillion, usually in evening dress as they would be staying for dinner. I was also taken, on two occasions, to the Passara Gun Club, to attend meetings of the Passara and Lunugala District Planters Association. There were also the odd shooting expeditions to the East Coast. These I enjoyed immensly. The P.A. meetings at Passara used to attract some great characters. There would be "Klondyke" Jamieson from far off Uva Estate at the end of the Madulsima range, old 'Tamby' Traill, George Kent Deaker of Gonakelle, Aubrey Clarke of El Teb, Maesmore Morris of Blarneywatte, and many others long since gone for good. I forget the name of the vicar who held a monthly service at the little church of St. Peter, in Lunugala. He usually stayed the night at Yapame when he came over from Badulla for a service. Services at Passara were held at the little church of St. Barnabas, situated close to the Passara Gun Club.

The educated Sinhalese then in the Lunugala District comprised, as far as I can remember, of the Sinhalese doctor at the Government Hospital and his family, and his Burgher counterpart at the estate hospital on Hopton. In Lunugala village there was a Transport Contractor and a few small 'caddai' (shop) keepers. The rest of the inhabitants were simple peasants who tilled their small plots of paddy land in season. A few grew stunted tea or coffee bushes and sold the leaf to the estates, or collected the fruit from their jak, breadfruit, plantain, or kitul and areca palms, which grew untended around their thatched houses. For most of the time they lived an indolent life. The Sinhalese villagers are physically a fine lot, well built and finely muscled, and their women folk can be very lovely with superb figures and carriage. They prefer gossip to anything else, and do the minimum of work necessary to support the life of the family. In fact the Sinhalese peasant is one of the laziest men alive as the poet referred to them in the following lines:-

"..................... are charming folk,
Polite and pleased to laugh and joke.
Though it is not to be denied
A trifle prone to homicide.
But one thing does make many boil,
It is their attitude to toil.
They do not think they were designed
For work of any wearing kind.

The Indian on the other hand,
Who settle in this lotus land,
Belong to some more busy breed
And labour very well indeed........
The Tamil plucks and prunes the tea
Which means so much to you and me."

Early in 1937 I was offered a job as SD to D.E.Hamilton, proprietor of Oodoowerre Estate, Demodera, at a salary of Rs.250 a month (about £.19), over half of which would be deducted to pay for my keep. I would live, at least for starters, at the Hamiltons bungalow. This seemed like riches to me and I accepted with alacrity, and with relief and gratitude that I would at last be starting on a career as a pukka Tea Planter.

CHAPTER TWO

WIRE SHOOTS AND AERIAL ROPEWAYS

The first Wire Shoot I ever saw was on 6th.November 1935. I remember the date clearly because the day before, at dawn, I had disembarked from the S.S. "Oronsay', in Colombo harbour at the end of my first voyage to the East.

This was the first of many voyages, but the only one I travelled by Orient Line, in which my father had been captain of the old "Oronsay" before the first world war.

The same day I caught the Night Mail to Badulla from Colombo's Fort station. We left in the noise and confusion, and humid heat of the late evening. The train rattled and bounced across the hot low-country paddy fields, through groves of coconut palms and endless lines of rubber trees, towards the central mountain ranges. Very early in the morning we woke to a chill dawn, the sound of rushing mountain streams, the hollow thunder of wheels crossing girder bridges over chasms, the puffing of the two engines hauling the train up the incline, and the rattles and booms as we charged into and out of numerous tunnels.

Eventually the train reached the summit tunnel at 6,300 ft and began its long descent towards the Eastern plains.

By this time I had donned a thick pullover and spent the time peering from the window at the magnificent scene set out below and away in the distance, where rugged purple hills reared into the sky. The engines and front coaches kept appearing to the right, and then to the left, as the train plunged downward round the many curved slopes at a speed of at least 20 miles per hour. Occasionally we stopped briefly at tiny rural stations where the guard and enginedriver chatted with the station staff, and Tamil and Sinhalese estate workers and villagers got on and off the train. The air was clear fresh and balmy. The only noises were the voices of travellers and the sound of the birds in the trees.

Just before we reached one such rural station, Idalgashina, I heard a rushing or swishing sound and something flashed through the air past my carriage window. I soon saw that it was a man, a

European, suspended from a long wire which seemed to disappear upwards towards a mountain top from whence he had come.

The wire terminated close to the station and as the train came to a halt the man, clad in blue shorts and shirt,and carrying a small suitcase, jumped into my compartment and gave me a cheery "good morning". My companion turned out to be an assistant on Idalgashina Estate. His bungalow was located high up on the slope above the railway, and he was setting out for the railway terminal town of Badulla, some 40 miles to the East and 3,000 ft. down. It was a Saturday morning and he was on his way to play rugger for Uva, the best, most rugged, and one of the most isolated planting districts, inhabited by planters known as the Merry Men of Uva. Little did I know then that I would spend the whole of my planting career of 37 years in Uva.

After the rugger match, and probably a celebration at the club, my new friend would catch the returning night mail, that evening or Sunday night and would descend at Idalgashina to be faced with a long climb up to his bungalow by the light of a hurricane lantern.

That was my introduction to wire shoots.

Soon I would come into much closer contact with this simple but effective means of transport.

Aerial Ropeways and wireshoots were used on many Ceylon Tea Estates owing to the difficulty and high cost of building roads in the steep and hilly terrain. Spring Valley, where I would spend well over 30 years, was exceptionally steep, even by standards in the Darjeeling district, and was broken up by gorges, valleys, precipices and mountain spurs. So it was many years before a road system was developed, Indeed, many parts of the property are still unroaded and never likely to see a motor vehicle.

Here a few technical details are given of systems of wire rope transport which were operated on estates where I worked;

1. The use of one Fixed Rope placed on an incline on which carriers (from which are suspended loads), are allowed to run down at high speed. This is generally called a "shoot" and it was one of these that I had seen at Idalgashina on arrival in the island.

The carriers for the loads were usually Michies Patent Wire Shoot Runners, which are fitted with an enlarged axle. In the body of the axle is a cavity for holding oil. This is filled when the runner is placed on the rope ready to start off down hill. The load is attached to the hook below the carrier wheel. Green tea leaf must reach the factory as soon as possible after plucking in an undamaged and cool condition. This necessitates that the shoot runners rushing down the wire (usually with a 70 to 100 pound load), be slowed down to arrive gently at the terminal rather than crashing into the ground. Planters have devised simple but effective mechanisms to achieve this.

2. The Single Fixed Rope, in which one carrier is drawn to and fro, hanging from a pulley wheel on the fixed rope, by means of an endless hauling rope which is usually motivated by a small engine sited at the upper end. More details of one of these will be given later.

3. The use of Two Fixed Ropes, with an endless hauling rope, in which carriers travel in one direction while others run on a parallel rope in the opposite direction. This is a thoroughly serviceable type of tramway capable of being used over extremely long spans, and carrying loads up

to 5 tons. This can be operated by electric or engine power. In the case of ours at Spring Valley it was motivated by electricity generated from our own Hydro electric power plant.

The wire shoots needed only the force of gravity to operate them.

Some of the advantages of this method of transport were;

a. Small initial cost compared with building roads and bridges and providing motor lorries.
b. Extreme symplicity and very low cost of working.
c. Ability to transport materials in a direct line over precipitous ground, rivers and defiles etc.
d. Small consumption of power when compared with tonnage transportable.
e. Small demand for labour in loading and receiving.

Disadvantages included lack of flexibility of points of departure and destination. In modern times efficient supervision of an estate requires access to all areas by at least light motorable roads, which can also be used - albeit more slowly - to transport material into and out of the estate, When I left the island in 1971 wire shoots and aerial ropeways on many estates were gradually falling into disuse. Spring Valley remained an exception for reasons already given, but even here some had been discontinued. It is sad that this "environmentally friendly" type of transport is becoming scarce.

My journey in November 1935 took me to Yapame Estate in the remote Lunugala District of Uva, where I started as a Creeper (learner) under my uncle "Flabby" Fowler. Yapame was mainly located at between 2,000 and 3,000 feet above sea level, but there was one Division which stretched upwards to near 5,000 ft along the sides of a deep ravine. The ravine contained a fine waterfall several hundred feet high.

Near the upper end of the ravine, on a tiny plateau, were situated the labourers lines, the conductor's bungalow, muster ground, with leaf and tool sheds, and Hindu temple. Close to the muster ground was a wireshoot terminal. The shoot ran across the ravine and down to the end of the estate cart road, from which point goods were transported to the factory by bullock cart. One day, whilst I was having lunch with my uncle and aunt a runner brought a message from the upper division's conductor saying there had been a serious accident and that a labourer, he named a popular member of the work force, had fallen to his death in the ravine. We hurried, uncle on his horse and I following on foot, to the scene about two miles away and up a precipitous track.

It turned out that Ramasamy, on a previous occasion unbeknown to the management, had himself lowered along the wireshoot in a basket to clear a jammed carrier and its sacks of tea leaf. This he had succeeded in doing and he was acclaimed as a hero by the rest of the labour force. He had indeed been very brave - or foolhardy - but broke strict orders that in the event of a jam occurring the wire must be lowered by the terminal operators at the winch.

On this occasion Ramasamy repeated his performance, but on being lowered to the jammed sacks - which were stalled above the deepest part of the ravine- he was unable to release them. He called to be hauled back up and taking an alavangoe (crowbar), was once more lowered to the obstruction where he tried to lever the carrier free. His efforts set the cable swaying and vibrating so that suddenly his own runner, attached to his basket, jumped off the wire and he fell to his death far below.

In due course I finished creeping and after 18 months on Oodoowerre Estate, I was appointed to the temporary charge of the two topDivisions of Spring Valley Group of estates under Colonel F.I.S. (Jock) Sutherland O.B.E., M.C., E.D. I was never confirmed in this appointment but remained on Spring Valley for well over 30 years and after 15 years took over the mantle of Jock.

There was a strict rule on Spring Valley, often re-iterated, that only specially trained workers were to operate wireshoots. In spite of this on one occasion during the war years, when supervision was short, a woman plucker placed a load of tea leaf on a mile long shoot to the old tea factory. She had the usual brass bangles on her arms. These or one of them, caught in the open weave hessian tea leaf bag and she was carried off down the shoot clutching desperately to the sack, her weight caused the wire cable to sag below the height of the lower terminal and she came to a halt a couple of hundred yards from the terminal, over a ravine. In spite of desperate efforts to lower the wire by means of the winch near the factory the woman's strength gave out and she fell onto the rocks below.

According to an estate map drawn in the early 1950's, Spring Valley had 17 permanent shoots for transporting green tea leaf towards its two factories by means of gravitation. As mentioned earlier there was also one Single haulage rope and an aerial tramway which were capable of hauling multipurpose loads upwards as well as downwards. The shoots were used three times daily when-ever those particular fields were getting their weekly plucking. Each workers basket is weighed at least three times a day, and the contents bagged and despatched to the factories. At the end of the day a pluckers total leaf weight is logged into a check roll. and he or she is credited with the weight which is paid for in addition to the daily wage.

On poor land, and land too steep and rocky for growing tea, fuel coupes are established - usually a variety of Eucalyptus, or Grevillea robusta, or perhaps toona (cedrella serrata) or Sapu (Michelia Champaca), etc. These supplied our own home grown timber for firewood and for building purposes. These coupes grew remarkably quickly, and when ready for felling or coppicing, tempo-rary wireshoots were erected to remove the wood to the factories and workshops, or for storage, treatment and seasoning for future use. The felled trees would recover to give a second and third crop over ten year spans. All well organised estates had these timber areas so as to avoid, as far as possible, reliance on imported diesel fuel for factories, and to provide our own building timber free of any purchase cost. We also quarried our own rock for roads and buildings, collected our own sand, and made our own cement blocks. These too were sometimes transported by shoots and ropeways.

The estate's Single Fixed Haulage ropeway had been erected in the 1880's and was operated by a 7h.p. Tangye oil engine. This engine was still working satisfactorily when I left the island in 1971 and, for all I know may still be in use in 1990, 110 years on!
It is sited in a small corrugated iron shed high up on the side of a rocky spur on Namunukula mountain. To get to this area on a small plateau with about 30 acres of tea - known as Korungu Mallay (monkey field), you had to climb 1,400 feet by zig zag paths. I often contemplated being hauled up on the wire rope. However breakdowns did sometimes occur, and the sight of a load hang-ing motionless, high in the sky, was enough to soften my resolve. There was a planter, Donald Hamilton who, in the 1920's used to be hauled up on this wire. I consoled myself that then the machinery and cables were only about 40 years old, not 60 to 80 years as in my time.

I knew Donald Hamilton well. He left Spring Valley, ran off with a wealthy wife of another planter, bought the fine Oodoowerre estate in the same district, which he expanded by opening several hundred acres of new tea clearings. For this he 'crimped' a large number of the best Spring Valley workers. My first paid job after creeping on Yapame was under Hamilton. In the autumn of 1937 he went on six months holiday to Britain with his wife and daughter, leaving me in full charge, at the age of 19, of his 1212 acres property. I lived in state in his fine bungalow with the house servants to tend to my needs. There was a large swimming pool, rare in those days, and the stables contained 3 polo ponies, two children's ponies, and there were 3 dogs, two milking cows, three cars a Chauffeur and syces. I did not have much time to enjoy these amenities as I had my work cut out in running the estate with all its ramifications. I did ride the ponies on my rounds of the estate and made full use of the nearly new Austin 7 tourer when attending Planters Association meetings and some social occasions. The only supervision I had was a fortnightly visit from a reputedly cantankerous old planter named 'Tup' Grant-Cook. Fortunately a friend gave me a tip that all would be well provided I produced Dutton's Dark Ale at lunch time and offered to turn out for the local cricket team at week-ends. This magic formula worked and I became very fond of the old boy who gave me much fatherly advice.

On his own property, Sarnia, Tup had a fine assistant by the name of R.A.Robin. Robin was a keen and expert rugger player, who was capped several times for Ceylon, but he was not interested in cricket. This greatly annoyed Tup who could not understand how any Englishman could be uninterested in cricket. The fact that Robin was a Scot simply did not enter into the matter. The outcome was that Robin was not allowed any leave out of the estate until he consented to subscribe to 'The Cricketer' and to turn up to bowl in the nets at the club. In those days some of the old planters were despots, but they had been through very hard times themselves and were nearly all men of character. They believed - quite rightly I think - that hard work has much to do with character forming, and provided we young fellows worked really hard generally let us play hard too, and invariably backed us up in emergencies.

But back to shoots and ropeways.

As we gradually extended the system of light lorry roads over the Spring Valley estates some of the old haulage ropeways were dismantled and removed. By the time I became Manager in 1955 we had only the Korungu mallay (monkey field) apparatus and a newish Dual rope Aerial Tramway for hauling loads upwards. The latter was over 1,350 yards in length, rising from 2,900 feet elevation to 4,800 ft. It was supported at strategic points on 6 steel pylons and was capable of carrying all types of materials up to, or down from, the New Factory at 2m.p.h., with 2cwt. loads spaced at 100 yard intervals. However transport on this tramway could be interrupted by high winds. The road with its nine hairpin bends covered the same destinations in 2 3/4 miles. Generally this tramway gave excellent service from 1948 until 1970. I believe there was only one minor accident during its operation. Unfortunately the next door property had a very nasty affair when the clothing of one of the operators got caught in a terminal wheel and he was dragged into the machinery before it could be stopped. That sadly was the end of him.

Siting and erecting wireshoots was often good fun. After deciding on the terminal points, and ensuring that there were no insuperable obstacles in between, the wire cable was transported to the bottom on its drum. Whenever possible this was done by lorry or bullock cart but sometimes man-handling was necessary. The top terminal was then clearly flagged so as to be visible from below. Markers were also placed on hillocks on the line between.

A gang of workers began hauling out the wire singing as they worked. Light ropes would be attached to the end of the wire to haul it over streams and chasms. Branches of trees would be lopped where these impeded the way. Eventually, sometimes after several days work, the wire would reach the upper terminal. Here a masonary or concrete plinth would be constructed, into which would be sunk long shafted holding-bolts for securing the cable. At the same time the bottom end of the wire would be attached to a large winch which would also be anchored to holding-bolts sunk in concrete or drilled into solid rock. When all was ready, with the concrete or masonry well set,the wire would be tightened by operating the winch.

Next, in the case of shoots for firewood, timber, or building materials, a tall timber barricade would be positioned to prevent the descending loads from smashing into the winch. In the case of shoots for green tea leaf an escape wire with a brake would be fixed. The shoot should now be ready for operation, a supply of shoot runners or pulley wheels would be taken to the top, and various loads sent down on trial and the winch used to adjust the tension of the cable. Finally the shoot would be judged safe and in order for daily use.

The estate blacksmiths were important people for this type of work. They and some other specialist workers had been taught how to splice with wire. They were skillful at carrying out repairs and received extra pay for doing this valuable task.

Communication between operators at the two terminals was essential. In the case of the larger aerial ropeways this was by telehone, the wires for this ran alongside the ropeway cables. In the case of wire shoots a system of signals was agreed by banging the cable. So many bangs for "are you ready for my first load?" etc etc. The workers were also able to communicate over considerable distances by voice, using yodelling like calls which carried very well through the hills and valleys, and which sounded very attractive and musical.

A special memory is of three wireshoots, arranged in descending order from high above my office and bungalow, to bring leaf zigzaging down across the valley, each shoot stretching from one side to the other. In monsoon weather, when the slopes were covered in cloud and visibility was less than in a London fog, the shoot men could be heard making their high pitched "hoo" cries back and forth, loud and clear nearby, but only faintly from far above and away below. This would be followed by the thrumming of the cable, then the hissing of the shoot runner. Sometimes the dim forms of the descending loads could be seen emerging from the mist, then being again enveloped in the swirling greyness.

Work went on throughout the year in every type of weather, from very hot and dry to soaking wet and cold and windy, when we would all return to our homes to warm and dry ourselves by log fires, and be ready to start work again before dawn next day.

CHAPTER THREE.

HOW IT ALL BEGAN

To the Secretary of State's question "Who are these coffee planters who are beginning to agitate in Ceylon?' The answer might be "A turbulent hill tribe, probably of Bohemian or Bulgarian origin, constantly on the borders of insurrection, but showing vague signs of a crude civilisation by hankering after roads, bridges and other visionary impossibilities".

We have run ahead of chronological events, but before returning to 1937 when Donald Hamilton offered me the appointment of junior assistant on Oodoowerre, it seems an appropriate moment to take a quick glance at the history of Ceylon's plantation industry.

The first coffee estate in Ceylon (excluding Hangurunkettia, the property of Mr. Charles de Soysa where some coffee trees are said to have been brought over from Arabia to grow flowers for Temple offerings), was opened in 1821 at Sinnapitiya near Gampola, under the management of George Bird. This land proved unsuitable and Sir Edward Barnes, the Governor, persuaded Mr. Bird to open land at Kondasale. Land was also opened at Black Forest near Pussellawa, so named because of the density of the jungle.

At first there was great difficulty in getting investors to purchase land for development as many thought heavy taxes would be levied on the produce. So the governor himself bought land and put in a manager to open it in coffee. He also encouraged government servants to do likewise, so as to develop the country. Kandy was the centre, and with encouragement from the top the demand for land rapidly increased, and all the surrounding districts were soon penetrated by the pioneers. There were no roads or bridges, life was hard and food scarce, but the opportunities for making fortunes in coffee were soon realised. The original settlers seemed to come out mainly from Scotland. They lived in huts with talipot palm roofs. The first shingles for roofing were not made until 1858, on buildings at Delta Estate, Pupuressa.

From 1821 to 1845 the coffee industry flourished. The Pioneers story is one of individual enterprise, hardship, fortune and heartbreak which could fill volumes. The tally of Crown Land sales to Ceylonese and Europeans grew enormously. From 3,661 acres in 1837 the total reached 295,526 acres over the next nine years. Major Thomas Skinner, the famous roadmaker and surveyor, drew the governor's attention to the vast expanse of uninhabited jungle in the central highlands. The land was surveyed and suitable areas marked out and sold. These blocks can still be seen on today's ordinance survey maps; recognisable by the straight lines drawn across the map, dividing the various plots which were given their individual estate names which many of us old planters know so well. The boundary lines completely ignore the rugged topographical features of the country. Hardly a corner of the vast forest regions of the Western slopes remained untouched. Into the valleys of Dumbara, Ambegamuwa, Kotmale and Pussellawa pushed the flood of settlers, some genuine planters, and some speculators who are always to be found in such ventures. They streamed into the mountain passes and spilled over onto the Eastern side of the island's central ranges, on to the rolling grass lands of Uva. Lands previously undisturbed except by the trumpeting of the elephant, the cough of the leopard and the bark of the deer were soon converted into fields of coffee. Sir Emerson Tennant compared this period of land development with the gold rushes in California and Australia, but with a significant difference that "the enthusiasts in Ceylon instead of thronging to disinter, were hurrying to bury their gold". This was a reference to the disaster which overtook Ceylon in 1845 due to a serious financial crisis in Great Britain. The government there abolished the preferential duty on coffee. The price immediately dropped from 105 shillings per cwt. to 48 shillings. As a result very few estates could make ends meet. Panic and consternation overran Ceylon. More than 400 out of 1,500 planters left the island to seek employment elsewhere. Estates were abandoned. Some idea of the seriousness of the collapse can be seen in the record of the sale of many valuable estates for a mere song. As an example; Narangalla Estate in Uva, which had cost £10,000, was sold for £350. Eventually, after strong representations had been made to the government at Home, the preferential tariff was restored and the island's coffee prospered again until the bushes were attacked by the microscopic fungus disease "hemileia Vastatrix". This fungus destroyed the leaves of the plant, and within ten years following its discovery in 1869 the industry was utterly ruined.

The catastrophy of 1845 to 47 resulted in a return to sanity and a laborious rebuilding of the industry. The planters who set about revitalising the estates realised that the many problems that faced them necessitated some form of Association be formed so that the community of planters could speak with one voice. Quite apart from the problems of finance, some proper organisation was needed to deal with the recruitment of workers, their transport to the plantations from South India. Complexities regarding workers housing and health needed sorting out, and there were many other aspects of the labour question needing attention. Since estate produce had to be transported over difficult terrain to the ports of Colombo and Galle, communications were of vital importance. These needed urgent improvement. There were many other problems requiring expert advice or concerted action for the common benefit. The initiative was taken by leading planters, and in 1854 the Planters Association of Ceylon was founded.

Thanks to the foresight and guidance of men of solid worth and wisdom, the P.A. was to prove influential in the development of the country. It was the planters themselves who went to the government in 1856 and asked to be taxed so that a railway could be built; and it was thanks to the P.A's continuous urging the government to build extensions and link lines, that a fine rail network eventually emerged. Later the P.A was equally involved with the development of the road system. It was a planter, R.S. Fraser of Matale, who installed the first electric telephone in 1880. The government was still relying on pigeons. Few people realise that to all intents and purposes a fair proportion of the provincial hospitals which exist today were built, equipped and staffed, at the instigation and expense of the coffee planters. A medical cess is still collected from estates by government, though the hospitals are used largely by the indigenous population free of charge, whereas estates have to pay for the treatment of their labourers. At least this was the position up to nationalisation. As will be seen in another chapter it was the planters, through the P.A., who took the initiative to tackle, and eventually eliminate, the menace of malaria. Almost every estate now has its own school or schools, and these are largely run and controlled by government. However it was Mr. Kingsford, the then P.A.chairman, who in 1904 advocated the establishment of schools on estates. Actually a few estates already ran their own schools. The first such was founded on Spring Valley in 1870, and I was on that estate at the time of its centenary in 1970.

These are a few of the very many ways the Planters Association influenced the development and prosperity of Ceylon.

It was Major Skinner (already mentioned as influencing the opening of coffee land), who pioneered the recruitment of Tamil labour from South India to work on the new coffee estates. In August 1840 he wrote to Governor Barnes from Abeegamuwa; "Who can view this perfect scenery without feeling it would be conferring a blessing on humanity to be the means of removing some 20,000 of the panting, half famished creatures from the burning plains of South India to such comparative paradise; benefitting not only them, the colony, and the individual by means of whose capital they would be brought here, but also our own native Singhalese people inhabiting the margin of this wilderness. Many totally unable to cultivate a grain of paddy .... would find themselves attracted to a new centre within this trackless wilderness,which (although I have often been jeered at for saying it), is destined 'ere long to become the garden of Ceylon; such a garden as has not entered into the minds of us pioneers to concieve",

Skinner's ambition and his prophesy were more than fulfilled over the next century.

The financial houses of Great Britain were taking a keen interest in the purchase of land for coffee, and the chief local bank, The Oriental Bank, gave its willing support to finance estates. It took almost a year from the time the crop was harvested until it was sold in London, so finance was all important. Loans were generously made against estimated crops. All went well until the onset of the new coffee disease. Meantime, by the late 1860s England was experiencing a great industrial boom. Money was plentiful but life in industry, or in an office, did not appeal to many who sought a life in the open air. Unfortunately English agriculture was in recession and openings in the Navy and Army were so keenly competed for that only a small proportion of the applicants were taken on. So many of those rejected were looking for work abroad. Ceylon seemed just right. These new entrants to planting life were rather scoffed at by the pioneers who maintained that they worked in kid gloves.

The newcomers were soon to receive a shock. Plantation coffee exports reached a peak in 1870 with a total of 1,054,030 cwts, but by 1889 had fallen to less than 100,000 cwts, thanks to "hemileia Vastatrix". It is impossible to convey here a picture of the misery brought about by this ruinous affliction. Fortunes were lost, the Oriental Bank collapsed in ruin. once more depression reigned everywhere. The planters were faced with the alternatives of abandoning their estates and the island, or sticking it out without funds and trying, somehow, to find new products quickly. Many crops were tried, and much misery endured before tea came into its own. The continuation of work on the estates was achieved in various ways. Some planters managed to take on partners who had capital. Others sold their properties to companies, either in London or in Ceylon, and took shares in payment: sometimes continuing to work the estates as paid managers to the companies. Although so many of the properties changed hands the original boundaries remained much the same, and still are the same, though the acreage under cultivation usually increased as the years passed by.

The decision to go for tea was an epic of commercial enterprise, skill and courage, and resulted in prosperity for the island for a hundred years. Within five years of the collapse of coffee 150,000 acres of tea had been planted. This was a great act of faith since tea brings no profitable return for at least four years from planting. By the end of the ninteenth century there were 384,000 acres under tea. By 1948, when Ceylon achieved independence there were approximately 550,000 acres of tea in the island, a total which has since fallen. Arthur Conan Doyle wrote of this remarkable feat of determination: "Not often is it that men have the heart, when their one great industry is withered, to rear up in a few years another as rich to take its place, and the tea fields of Ceylon are as true a monument to courage as is the Lion at Waterloo".

On the Uva side of the country the estates were much more isolated and thinner on the ground, and only comparatively small areas of the properties had been opened up in coffee. The result was that when coffee failed the estates usually had plenty of uncultivated land still available for opening in tea or cinchona, or whatever crop was considered suitable. Oodoowerre was one such estate. The first coffee estate in Uva was planted at Ridipane near Badulla by Major Rogers, the Assistant Government Agent (of whom more later), in the 1830s. This small estate was such a success that one of Roger's assistants, Assistant Surgeon Galland, bought and opened a property called Maryland (now called Kottagodde and part of Spring Valley). He also bought another property, Oodoowerre, in partnership with his brother-in-law A. Bertlin. In 1884 Oodoowerre was marked as "owned by the heirs of A. Bertlin" and, at that time, consisted of 685 acres of which 393 were cultivated. The estate was then under the charge of H.O.Hoseason, a great athlete and horseman, and keen on all sport. He later became very well known as Manager of Demodera Group, one of the largest estates in Uva.

It was said that as a host no-one could surpass him, and so many were his guests at Oodoowerre that he built very large stables to accommodate the horses of his visitors. Later Oodoowerre became the property of Packir Saibo, a merchant of Badulla. D.E.Hamilton and a partner named Nicholson bought the estate from him soon after the Great War of 1914 - 1918.

Hamilton was the dominant partner and put in a great deal of work and money to improve the property. He increased the acreage by buying more land, and opened up patna areas, at around 4,000 feet elevation, in tea. A modern factory and a fine bungalow were also built at this elevation, with a large swimming pool (a novelty in those days), stables for six horses, and garages for several cars, not to mention staff and labourers housing, workshops and all the other erections necessary for the working of a tea estate. A steep approach road with many hairpin bends had to be cut, to reach the new sections from the government 'cart road' in the valley below. This extended for three miles to the bungalow, and then continued to spread out so as to reach the factory and all the working areas.

At that time I knew nothing of the history of Oodoowerre, and very little about the background of the industry, nor did I appreciate that my new billet offered a promising opportunity on a lower rung of the ladder of advancement for a young planter.

Scarab Beetle rolling a ball of cow dung

CHAPTER FOUR

ODOOWERRE

"Fate, Time, Occasion, Chance, and Change ?
To these All things are subject........"

Shelley "Prometheus Unbound".

D.E.Hamilton, the proprietor of Oodoowerre, had invited me to stay on two occasions before offering me a job. I was not aware that he was looking for an S.D. In fact I learned long afterwards that he was looking for a young man to train, over some years, to take charge of the whole property and manage it for him. He wanted to devote much more time to his many other interests and ambitions. He hoped to become, and did, the Chairman of the Planters Association of Ceylon. He wanted to represent the tea industry in the State Council, as a nominated European Member. This he failed to achieve. He was already the first Chairman of the newly formed Ceylon Planters Society. This last named body was formed by popular demand of planters, to look after their interests now that the vast majority were paid employees with no financial interest in the properties they superintended other than their emoluments. Individual estate proprietors had become a rare breed, most estates being formed into public companies. At the elections following the granting of universal sufferage to Ceylon in 1930, under the Donoughmore Commission, Donald Hamilton had stood as a candidate. He was defeated, but not by a very large margin, which was surprising as he was standing against a well known and popular Sinhalese. He realised that if he was to be successful with his ambitions he would have to devote more time to them. Hence his need for training up someone to manage his property.

So when I was invited to Oodoowerre I thought it was just because the Fowlers and Hamiltons were friends and I treated the visits as purely social.

Early one morning during the summer of 1936 I caught a rickety old bus at the old Yapame Store. As an European I was invited to take a front seat alongside the driver and we set off on the 25 mile drive to Badulla. In those days, before the founding of the Ceylon Transport Board, the buses were privately run, usually on a shoe string. They had open sides and bench seats right across the width of the vehicle. Passengers simply climbed up the side wherever there was a vacant place on the bench. If there was none they hung on outside. The various owners competed madly with each other to attract passengers. To save petrol they invariably free-wheeled down hill. Tyres were usually bald and nearly open to the inner tube, so that punctures were commonplace, not to mention the risk of going over the khud.

After a short stop in Passara town where we drove up and down the single street touting for passengers, followed by a rival on the same errand, we climbed to the Debedde Gap at about 4,000ft. elevation, still hotly pursued by the equally decrepit bus of the rival owner. We then proceeded to freewheel down the 2,000 foot descent to Badulla, twisting and turning round the hairpins and the precipitous sides of the mountains into whose side the road is cut. In those days the road had not been widened and there was scarcely room for vehicles to pass in opposite directions, so the risk of crashing into another vehicle was considerable, especially as the conductor was shouting a running commentary regarding the progress of our pursuer who was close behind.

At Badulla, which I reached with much relief, I took another bus. Again we toured the town touting for passengers and did not set off on the Bandarawela road until we had a full complement. This time however there was no rival and we travelled along a reasonably flat valley to Haliela and then onwards to the old Oodoowerre Store, where I alighted and made my way to the old bungalow. This was built by Hoseason in the coffee days as described in another chapter. It was now partly office and partly a residence for a retired bank manager from the Bank of Uva, and his wife and daughter (whose names I forget), by the grace and favour of Mr. Hamilton. A message was sent to the Hamiltons in the new bungalow situated far above, near the top of the estate and Joseph, the driver, was sent down to fetch me in the Standard. I remember very little about this first visit, except that I met, for the first time, Mr. and Mrs. Hamilton, their eight year old daughter Eve, her forbidding European nurse Mrs. Crosby, widow of a sergeant major, and the plump, jovial Commander A. K. Hallilay RN Retd; the new Secretary of the C.P.S. Also present was a horsey young man with blonde handlebar moustaches, named J.Mansergh Hodgson, who told me he was "on the cart road", but being employed temporarily by Mrs. Hamilton, helping her organising soup kitchens in Badulla, to feed the victims of the malaria epidemic amongst whom were quite a number of orphaned children. Being "on the cart road" was a very serious matter. There was no financial assistance of any sort. You either packed up and sold up hoping you could raise enough money to get a steerage passage Home, or found some friend or philanthrophic individual to live with until better times. Mansergh was of the latter catagory and Mrs. Hamilton was a very kind hearted person as I was to discover for myself.

I paid a second visit to the Hamiltons later that year. Afterwards, on 26th September 1936 I wrote to my parents telling them;-

"I had a simply topping time. I arrived there on Friday morning and there was a girl named Gillian staying.(I remember nothing whatsoever about her). In the afternoon we had a marvellous bathe in the private swimming pool which is perfectly situated in a little hollow 4,100 ft. above sea level, and from the lawn which surrounds it simply marvellous views can be seen.
Mr. Hamilton came to the island about 30 years ago without a cent. After 15 years he bought Oodoowerre, about 500 acres of tea. He gradually opened up over 200 acres of patna and jungle land, and now it is one of the best tea estates in Uva Province for its size, and he is the owner. There is a climb of over 2,000ft. up their own road to the bungalow which is a beauty. They have three old polo ponies which are now used as hacks, and two children's ponies. He has a Baby Austin tourer for estate work, and a very nice Standard, and a beautiful 8 seater Humber. Joseph their driver met me in Badulla with the Standard and drove me to the bungalow, 10 miles .

Mr. Hamilton is known as a bit of a Driver and says that all SDs ought to be up at 5am and sent to bed at 7.30pm, and that they don't do nearly enough work nowadays. I had a dip in the pool at 6am on Saturday and then we went in to Badulla for the tennis tournament at The Uva Gymkhana Club. Unfortunately my partner and I weren't much good, so we didn't put up much of a show. The two Pickering girls who were at Oodoowerre when I last went were there, and a girl called Betty Smythe, and a very nice girl named Phoebe Grant-Cook, and lots of other people. Before lunch I had played 26 games. We had a cold lunch at the club and afterwards carried on with the tennis (GOSH was it HOT!!), and we played 18 more games. Then there was a bike race for the natives, and a polo match, and after tea there was a treasure hunt, aunt sallies, Coconut shies etc. At 6.30pm Mr. and Mrs. Hamilton went home and a crowd of us went to the cinema (silent and made of corrugated iron, but good fun). There was a dance after the cinema but unfortunately my partner was so tired by then that she didn't want to go. So we returned to Oodoowerre and after I had changed I decided not to go in again. I wish I had though, as it was evidently great fun and didn't stop 'till 4 this morning." (my partner', was the daughter of the retired bank manager referred to recently. She was what was then known as "Country Bottled'; i.e. had been brought up and educated in the island, in fact she had never been "Home". The poor girl, whose name I forget, had an old soak for a father who apparently drank all his savings and pension away. She was a nice girl living in very difficult circumstances and I hope she is now well and living happily).

"I had another dip early this morning and then went in for the finish of the tennis and clay pigeon shooting. Then at 11.30 I had some cheese and biscuits and caught the smelly old bus, arriving here(Yapame) at 3.30 feeling famished. Uncle has gone over to Neluwa Estate, which he visits every month, and has taken Aunt Dollie with him, so I am alone now".

The old tea on Oodoowerre was nearly all down in the valley on old coffee land at 2,000 to 2,750 ft elevation, and was under the charge of Mr.V.Irlappen, an SD from Malayalum country in South India and a fine man. The younger tea, comprising about half the total acreage, had been planted far above at elevations of 4,000 to 4,500 ft.. The factory, also built by Hamilton, was situated in the latter area, which was to be my responsibility. Between the lower and the upper divisions was a hillside of several hundred acres covered in patna (a long coarse grass) The estate road climbed steeply up through this grassland twisting and turning through nine hairpin bends. The new bungalow was near the top hairpin.

Whether Hamilton decided I was just the man for the job or,more likely, there was no-one else available, I was offered and accepted the post in the spring of 1937. I was put to work as soon as I arrived; rising before dawn and attending morning muster of the labour force as the sun rose behind Namunukula mountain to the East. When crop was good the evening leaf weigh-up was often finished by the light of a hurricane lantern. Not only was it necessary to work long hours to get the crop in before it grew coarse, but it was difficult to pursuade the pluckers to stop work. They wanted to earn as much as possible by way of the poundage bonus which was offered. After completing weigh-up I would hurry back to the bungalow and write up the pocket check-roll in my room then, when it balanced, transfer the figures to the Big Check Roll which formed the final basis for the workers monthly wages. At the end of each month I would be up most of two or three nights, calculating the gross wages of each worker, then deducting the cost of advances such as rice, cumblies, other foodstuffs etc,-, and then working out everyone's monthly pay. Wages depended on the number of days worked, the type of work done (e.g. pruners and certain other workers were paid special rates and the quantity of green leaf brought in by each plucker was subject to a bonus payment).

Uncle Bernard Bean sketching amongst the mana grass at Cocowatte, 1937, when over from Bengal, staying at Yapame.

Namunukula Peak, 6,999 ft and Spring Valley from Oodoowerre bungalow. Swimming pool in the foreground.

About twice a week, sometimes more often, the Hamiltons had guests to dinner followed by bridge, which they often, played late into the night. Usually I had my meals with the Hamiltons but guest nights were an exception when I took my supper in my room so as to be free to complete my book work and spend some time studying to learn Tamil to pass the C.P.S. examination. However I used to be called in sometimes to make up a bridge four. These calls increased in frequency as time passed. Not only was I a complete novice, but was often far too tired, after a long day's work, to concentrate intelligently. The post mortems after play finished bored me to distraction and having been kept until after midnight I still had to be up for morning muster. I loathed these bridge sessions with people far older than myself and swore never to play the game for the rest of my life; nor have I.

Hamilton certainly kept my nose to the grind-stone, which probably did me no harm looking back on it. There was no time to take note of anything going on outside the estate. However, one day whilst checking the topmost boundary I met another planter checking his. He was Philip Grimwood the SD in charge of the Napier Division of Rookatenne Estate. He invited me to lunch the following Sunday and I borrowed Caesar, the best of the old polo ponies, to ride the three miles to his bungalow. I was envious of his living in his own quarters and the independence it allowed. After that we met quite regularly at week ends. Joseph was teaching me to drive on one of the old leaf lorries and I used to arrange for Sunday lessons or, if there was plucking that day, I would manage to arrange a detour for a lorry to drop me off at Napier on its way to collect its load. I believe that Grimwood later married the daugher of C. Carson-Parker of Shawlands Estate, Lunugala. After the war he went to Kenya and was decorated for bravery when he put up a spirited defence of his property against attacking Mau Mau terrorists, driving them off and killing several.

The three Oodoowerre lorries were built on the strengthened chassis of old eight cylinder Wolesley cars, and were powerful enough to haul loads up all the steep hills on the property. Joseph and his assistant mechanic did a fine job in keeping these old engines in trim. In addition to the main arteries through the estate, Hamilton got me to convert all the old riding roads and many footpaths to take his Baby Austin. He seldom walked but boasted that with this road system he could closely inspect every inch of his property. Some of these slightly enlarged paths were hair raising to drive on. A small error of judgement would have meant plunging hundreds of feet. Having learned to drive on such roads I had no difficulty in passing my driving test on the flat streets of Badulla where the only obstacles were pedestrians, cattle and stray Pie-dogs which roamed at will.

When it came to economies no-one could touch Hamilton. There was hell to pay if a tea bush was lost due to a corner being widened a fraction more than was necessary. Quite rightly (but forgotten by many planters), every tea bush was highly valued for its potential return. The culverts on the estate roads were made with empty 45 gallon oil drums after the oil had been used up at the factory. However later on there was a shortage of these barrels and they acquired considerable value. During the night villagers would creep up and dig them out with the result that care had to be taken when setting out in the morning. Nevertheless he had his priorities right. If something would bring a profitable return he would be prepared to put up the money.

Mention has been made in another chapter of Hamilton's early days on Spring Valley; how he travelled on the leaf ropeway, and crimped labour from that estate when he acquired Oodoowere. He made a great success of his estate and increased the yield dramatically to around 1,000 lbs made tea per acre, which was exceptional in those days. The estate had been allocated a generous quota under the International Tea Control Scheme. In addition Hamilton bought up quotas from several small adjoining properties. The control scheme proved such a success that it was abolished within less than five years from its introduction in 1933. Oodoowerie was blessed with a sound programme of fertiliser applications - generous for those times - and pruning cycles extended to four and five years for the vigorous young tea in the upper elevations. These policies were rather ridiculed by some of the old planters who felt the bushes would not stand such treatment for long. He was a great believer in combining generous applications of green stuff, mulch and compost to supplement the artificial fertilisers. All prunings, plus heavy crops of dadap (Erythrina lithosperma), gliriciddia (g.maculata), and Accacia (cupressus macrocarpa), loppings were forked in between the tea rows. Patna lands adjacant to the tea gave plentiful supplies of maana grass (cymbopogon confertiflorus), which was turned into compost for additional mulching.

Hamilton claimed to be the possessor of a photographic memory and that his filing system was kept in his head thus saving much time on paper work, as well as the cost of clerical staff. I was so impressed by his memory and powers of observation that I decided to take a course in the Pelmanism method of mind and memory training. I completed the course and the ability served me well for years but the books were lost during the war. They would have been useful in recent times.

In due course Grimwood was replaced at Napier by a man named George Bolster. One week-end the three of us took the train from Demodera to Ohiya station at the edge of the Horton Plains. These plains are situated at an elevation of 7,000 feet and above, and after the thick forest on their approaches come as a great surprise. Huge, treeless, rolling patnas, with small patches of jungle here and there in which can be found Sambur (Elk) and other wild life including leopard. The plains spread for miles and are often wreathed in thick mist. As the clear streams hold fine trout and stags graze on the patnas the plains are said to resemble the moors of the Scottish Highlands. Copses of juniper and rhododendron abound, as do tree ferns. If you happen to be there at the right time there is a wonderful display of nillu flowers (strobilanthus). They cover the ground in a massive carpet of red and blue. They are everywhere and attract masses and masses of wild bees as well as large numbers of jungle fowl which get drunk on the flowers. However the nillu flowers only once in seven years, and having bloomed they die.

From Ohiya station we climbed the steep jungle clad escarpment onto the plains, seeing a troop of Wanderoo monkeys travelling through the tree tops above us. In the highlands these monkeys grow to a great size and being covered with a thick layer of fur they seem almost a different breed from their cousins in the Low Country. From the top of the escarpment we followed the track to the Rest House which is situated more or less in the middle of the plains. Here a meet of Hounds was to be held, and about ten hunt followers had walked up with the hounds from the Agras. The meet was not very successful as the pack kept getting split up with sections following different scents. The idea was to hunt Sambur but some hounds got onto the scent of a leopard and went after it. They brought it to bay and there was a terrible fight in which one hound was killed and others wounded. The weather was very cold, with thick, damp, mist. Sounds coming out of the mist were difficult to locate and followers and huntsman gradually found their way back to the Rest House and its roaring log fire. I remember the bedroom was very cold and bare and we had to break the ice on the tin wash basins in the morning. 'Vandy' Van der Kiste had joined the hunt from the very isolated Nagrak Estate and had invited the three of us to to lunch the following day. We set off after breakfast, walking through the dripping moss clad jungle, and thick mist, for the 5 mile walk to his bungalow on the southern edge of the plains. He warned us to be careful not to fall over World's End, but we didn't even see a sign of that precipice which lies close to the path. Harry Williams, in his "Pearl of the East", describes World's End as "the sheer fall in the nature of 5,000 ft, but figures convey nothing of the wild fascination of the place or of its cold grandeur. Rivers and forests, plains and hills, stretch like a cunningly woven carpet to the delicate line of the horizon, rimmed by the sea. World's End is also the extreme limit of the Garden of Ravanna". We saw nothing. Vandy seemed to have an inexhaustable supply of champagne, though we must have made quite a dent in his stock. We left for the return journey late in the night, and stumbled our way back to the Rest House with the aid of hurricane lanterns, but still without a sign of World's End. I think we doubted its existence.

In those days there were no motor roads up to the Hortons. You could get there on foot or horse from the Agras, Bellihiloya or from Nuwara Eliya via Pattipola, and there was the steep foot track from Ohiya station.

In August 1937 I wrote Home remarking how marvellous it would have been if I could have joined the family at the Waldringfield Regatta. I said I could clearly imagine it all because it had been such fun during my last summer at Home in 1935, and apparently the new pram, the "Chick" to "Cape Pigeon", was no mean sailer by all accounts. Perhaps I was feeling home sick after nearly two years away. I continued:-

"When Hamilton returned here after his ten days holiday he made no criticisms, so I hope he thought all had gone well. He is away again for more meetings next week, so it is getting me used to being in charge. He will remain on the estate all the rest of September, and they all sail in the "Mooltan" during the first week of October. Before he goes he wants to get all the important work on the new clearing finished, so there is sure to be a bit of a scrimmage. There is any amount to do and only a month to do it in."

We had a couple of rows before he left. I thought he was being unfair both to me and to some of the staff, blaming us all for mistakes he had made. We had some very angry exchanges and I thought this boded ill for the future. However I was left in charge of the property with only "Tup", (S.E.Grant-Cook) visiting fortnightly from Sarnia Estate, an hour's drive.

I was very fortunate to be in a situation which ought to have been the envy of other young planters, few if any were as young as I. Fortune appeared to be favouring me, for with no particular qualification other than a good grounding as a creeper, and not yet aged 21, I had almost complete autonomy and charge of a fine estate with all the ramifications of its day-to-day organisation; even if it was to be for a limited period of six months. Fortunately the staff were excellent, and so was the labour force. I thoroughly enjoyed the responsibility,and the challenge to do a good job was the best motivation there could be. Most Indians, and this was confirmed later in the Indian Army, respond enthusiastically to any keen youngster even if their own experience is very much greater than his, and will stand by him and quietly and tactfully give their good advice. If, in turn, the youngster listens and responds there will be the beginning of a good and loyal relationship. Fortunately I had been taught f rom the outset to remember that I was a foreigner working in someone elses country, and must never forget that.

In those days many SDs were not even allowed inside their estate's tea factories until they had spent years in the field, and few expected to get an acting charge before their second furlough, a period of 8 or 9 years.

However I had been shaken by Hamilton's inconsiderate attitude in the weeks before he left. Had I seen the writing on the Wall? During the next couple of months I gave the matter quite a lot of thought. Oodoowerre was a one man show. Would I be happy for long under Hamilton? I was entirely in his hands with nobody else in authority to turn to. Would it be sensible to join a large company owning several estates, or to get into the employ of one of the reputable agency houses which supplied PDs and SDs to tea and rubber properties throughout the plantation districts? These thoughts evolved gradually as I got to know a bit more about the district and the numerous estates and their superintendents. I also sought the advice of experienced planters such as Flabby Fowler and Tup Grant-Cook. I made up my mind after a couple of months that I would start looking for a new appointment.

The North East Monsoon broke towards the end of October. For days on end the bungalow, factory and upper sections were enveloped in thick dripping mist. There was a great deal of rain. I would come in soaking wet morning and evening to a roaring log fire in the sitting room. One wet and misty evening the head servant came and told me that a 'lady' had arrived and wished to speak to me. When she was sent into the sitting room she introduced herself saying she had got lost in the mist. I wondered how she had managed to get off the main road onto our steep and narrow estate road, and to climb the nine hairpins before realising she was lost. She was middle aged (or seemed so to me), was smartly dressed and had an expensive car. She spoke with a foreign accent and claimed to be a German on her way to visit some fellow nationals at some destination I had never heard of. I gave her a drink and dinner and, as I was not attracted by her company, I telephoned a Swiss planter near Ella who, when I had explained the position kindly told me to direct her there so that his wife could entertain the lady. I do not rember the outcome of her visit except that some of my friends reckoned she was probably an expensive tart, and others thought she was a German spy. In either case she was at the wrong place and my finances could not support expensive entertainment.

After my 21st. birthday in February 1938 I wrote to my parents thanking them for their gift of a gold signet ring bearing the family crest and motto. I went on to write that;-

"I expect my 21st. birthday has been a bit different from most but I enjoyed it nevertheless. I was up at 5.30 and, after checking the thermometer under the tulip tree, had a dip in the pool before doing some exercises with my rifle and an alavangoe (crow bar) as dumbells ! Then after a quick break-fast of porridge and kedgeree I went to the factory. At 7 Ceasar was brought there by the horsekeeper and I rode to the end of the 75 acres with all the dogs accompanying me. It was very bright and my new sunglasses haven't yet come. After chasing 160 pluckers I went across to the 104 acres where there were 40 coolies forking in compost and green manure. I sent Caesar back and spent an hour there before going round some line rooms with the dispenser; then to the site of the new lines and pegged out more foundations. While there I discussed with the Baas the method of fixing the roofs. Then back to the factory where I tasted the previous day's make. I got back to the bungalow at noon and went through the incoming mail; valuations of tea shipped to London; a letter about opening a new road etc. After lunch at 12.45 I read in the newspaper about Mr. Eden's resignation, then drafted some letters for the clerk to type, before going to see the afternoon weighing. I got in a short time ago at 6.45pm. I hope actually to celebrate my birthday next week-end when Frink will probably be in Colombo".

(This was the family name for Francis, younger brother, and No.5 in the order of succession. I was No.4. He was then an apprentice with Alfred Holt's Blue Funnel Line, but on loan to M.V. "Glenapp" due to call at Colombo)."

"On Monday evening I went over to Sarnia and first of all went into some figures etc., about estate affairs with Mr. G-C, then down to his factory with him to see some new apparatus he has installed. I then went back to the bungalow and met Mrs. G-C and Phoebe. I had a very pleasant evening and was lent some nice books. When I came away MR.G-C said that I must play cricket again on Saturday and, if I can control my spin, I ought to be a very useful left arm bowler for the team. He also told me that Mr.Sentence-Smith, the Manager of Telbedde Estate, 1500 acres, wants an SD, and he has recommended me. Last year Telbedde paid a dividend of 30%. There are two other SDs. H.Williams and Hugh Breay". (Harry Williams was the author of 'Ceylon, Pearl of The East', and other books on Ceylon).

"This afternoon Major Sutherland of Spring Valley Estate, 2000 plus acres, rang me up and said he wants a temporary SD for Mai Mallay Division (5,000 - 6,000 ft). He didn't actually offer me the billet but wanted to know if I'd be free. It would be a very energetic time he said, so I imagine he is erecting buildings and opening roads etc. The present man who is going on leave is aged about 40 and is paid Rs.600 a month. He said I would be senior to the junior SD. I have not accepted anything and will wait until I have been to Colombo. It certainly looks as if Mr. Grant-Cook has been throwing his weight about! Several times Spring Valley has taken on men "temporarily" and has kept them permanently. The S.V. job would be much the best paid and far the nicest and healthiest billet".

I was told by Grimwood that Mr.Will of Glen Alpin Estate was asking about me. So it looks as if there is no dearth of billets. However I will see what they say in Colombo. I am sending a map of Ceylon, not a very good one. It shows the roads but no contours. No; the Horton Plains are NOT South East of Badulla but close to N'Eliya. On the map it looks as if it is near roads but actually it is cut off by steep escarpments. Often when roads look close together on the map they are separated by unclimbable precipices or unbridgeable gorges. Oodoowerre looks close alongside Demodera, but actually is 8 miles away by road, although you look right down on Demodera station from our factory 2,000 feet above.

I.have had nice long letters from all the rest of the family and wonder what their present is. Tup says he posted it so it should arrive soon".(Tup = Cuthbert, an elder brother, No. 3 in the line, commissioned with 2nd. Battalion, The Kings Shropshire Light Infantry).

"27th February. 'Glenapp' will arrive on 3rd March according to the paper. A letter has come from Major Sutherland saying that the Spring Valley job is really permanent. Mr. Grant-Cook advises me to accept it if it is offered to me.

Quite an exciting birthday week ! By the way; we beat N'Eliya yesterday by 15 runs. I made a rounded figure but took four wickets.

P.S. "Glenapp NOT calling COLOMBO". Thus a telegram from Whittalls just arrived".

It was very disappointing that the ship was diverted. When my brother reached Alexandria, being in the Royal Navy Reserve, he reported to the Naval Authorities owing to the International crisis.

He was appointed to a warship of the Mediterranean Fleet, but when the crisis eased he was released. After arriving in Britain he joined another merchant ship which called at Colombo later in the year when we had an enjoyable reunion.

During 1937 I joined the Ceylon Planters Rifle Corps, a volunteer unit of the Ceylon Defence Force based somewhat on the lines of the Territorial Army in Britain. The role of the unit was concerned with the defence of the colony and its internal security. The commanding officer was Lt. Colonel F.O. Sprinks, a planter from the Dimbula District. Major F.I.S. Sutherland M.C., was the second in command. Oodoowerre bungalow looks east towards the mass of Namunukula (nearly 7,000ft) and its nine peaks, and has a fine view of the great Spring Valley estate clinging to its slopes. Major Sutherland was the manager.

I was a member of No-9 platoon (2nd. Lieut Hugh Meldrum and Sgt. A.P. Fincher), of No.3 Company (Major "Scrubby" Lushington). The adjutant was Captain Taylor M.C. of the Rifle Brigade, who was supported by a fine team of regular warrent officers. Having been a sergeant in my school Officers Training Corps, and passed Certificate 'A', I was not entirely green as a rifleman recruit. Being taller than average I often found myself as right marker for ceremonial parades (tallest on the right and shortest on the left). All members of the corps were issued with uniform, equipment, rifle and sword (bayonet, to infantry of the line). My rifle and sword were numbered 796, and the bolt was No-1-86582. Corporal Budge Birkett, an SD on Sarnia under 'Tup' Grant-Cook, was my section commander and responsible for seeing that I kept my arms and equipment in proper order. He was also the platoon bugler and could sometimes be heard practising when I called on him at week-ends. Some of the section and platoon parades were boring and when it came to dis-mantling and assembling a Lewis gun it was a case of the blind leading the blind. However Company and battalion camps were organised by the professionals and were usually good fun. Even the section parades were well attended as leave for these was seldom refused. They often finished with a party at the Uva Gymkhana Club. With the international situation becoming more and more tense, with Germany, Italy and Japan posturing and threatening, we began to take training more seriously.

My first ceremonial parade was on Remembrance Day 1937. The platoon fell in outside the Uva Club and marched through the town followed by the be-medalled Old Comrades of the South African and the Great War (1914-18). At the War Memorial near the kaccheri we formed line and reversed arms for the two minutes silence,and the Last Post and Reveille, blown by Budge. The vicar of St.Mark's church took the service, which was well attended by the town's people and the planting community's families. It was a moving occasion in an outpost of Empire.

In 1937, or was it 1938, the batallion joined the Royal Navy at the naval base at Trincomalee for some exercises. The whole of the C.P.R.C. went by troop train from Colombo and, owing to a short-age of drinking water at Trinco, two water tanks were attached to the rear of the train. During the night the train stopped in the jungle, not far from Maho junction, and started to run backwards down a long hill. We had failed to reach the top of the rise owing to the heavy load. After backing well away onto the flat we rushed at the hill again., the engine puffing and chuffing away like mad. All the troops were awake by now and bets were taken as chuffing and puffing gradually became slower and slower. No; we could not reach the top. Once more we backed down onto the flat and a water truck was disconnected and we tried again, without avail. It was not until the second water truck had been discarded that we just made it to the crest and continued towards Trinco waterless. At Trinco the Royal Navy came to our rescue - with beer !

Caesar, Ladybird and Folly with J. M. Hodgson, Eve Hamilton and Mrs Crosby, at Oodoowerre, 1937.

No. 3 Company, Ceylon Planters Rifle Corps, at Diyatalawa, 1938.

Middle right and bottom: Demonstration use of the bayonet (The author and "Sandy" Richardson).

47

The C.P.R.C. had to defend the coast north and south of the naval base against landing parties from H.M.S. "Norfolk" and H.M.S. "Gloucester". I don't remember which party, if any, was declared the winner, but after the conclusion of the exercise we all boarded the ships for a voyage back to Colombo round the south coast of the island. We were messed by the matelots who were a very fine lot, but they just could not understand why we planters played at soldiers "for fun". The ranks of the C.P.R.C contained all sorts. An SD might be an officer or N.C.O., with his PD possibly a rifleman under him. Heads of Colombo business houses, and senior bankers or civil servants were often rankers, with a shop assistant (shoppie) perhaps being an officer or warrant officer over them. During the voyage a rifleman of the Colombo Company was severely reprimanded by a midshipman for some minor misdemeanor or infringement of the ship's rules. When the ship arrived at Colombo we all disembarked and, only about an hour later the Governor paid an official call on the ship. He was accompanied by a civilian A.D.C dressed in the magnificent uniform and cockaded hat of that office. . The goggle eyed midshipman on duty at the gangway immediately recognised the A.D.C as the rifleman he had so recently repremanded. The A.D.C's face remained expressionless except, perhaps, for the faintest of winks.

The other main units of the Ceylon Defence Force were the C.L.I. (Ceylon Light Infantry) another part time volunteer unit which was officered and manned mainly by Ceylonese. The C.G.A.(Ceylon Garrison Artillery), The C.E (Ceylon Engineers), and the C.M.R. (Ceylon Mounted Rifles). Like the C.P.R.C. the C.M.R. was recruited from amongst Europeans of the planting and commercial houses. All ranks had to supply their own mounts. This unit was amalgamated with the C.P.R.C. in 1939, when it provided a motorised recconnaisance section. Their own vehicles.

The C.M.R. generally provided a mounted escort for the State Drives of the Governor. One such State Drive was routine for Governor's Cup Day at the Colombo race course during the annual August Week races. The troop of C.M.R. providing the mounted escort positioned two files of three horsemen ahead of the open carriage. More files of three followed behind and other horsemen rode on the flanks. On one of the last Governor's Cup Day's before the war it happened that the trooper occupying the centre position of the leading file had celebrated too well in advance. His inebriated state was noted only by the troopers on his port and starboard sides after the cavalcade had set out. It was too late to make any change of plan. They were already on their way and drawing close to the race course. The Governor in his magnificent uniform, cockaded helmet, medals and orders shining on his chest, sat in the carriage with his fashionably dressed and behatted lady as the procession rolled majestically towards the Grandstand,to the cheers of the crowd and the music of the police band. Nobody seemed to notice the unsteady swaying of the trooper riding in the centre of the leading file. His chums on either side were riding stirrup to stirrup to help keep him upright.

Suddenly there was a loud explosion as an exuberant spectator let off a thunderflash. The horse, already upset by its riders uncertain gait, leaped ahead and bolted riderless down the course. At the first bend, seeing a gap amongst the crowd, it jumped the railings and disappeared from sight. But the inebriated rider's flankers rose to the occasion. Just in the nick of time they each put a hand under his armpits. Thus they conveyed their horseless companion, suspended between them still correctly positioned and trotting in front of His Excellency's carriage, to the end of the course and to the wild cheering of the crowd.

History does not relate what happened to the culprit afterwards. I always mean to ask him, but invariably forget, when I see him at annual reunions of the C.M.R and C.P.R.C. These are still held more than forty years after the units disbandment though sadly and inevitably the ranks are growing thinner and thinner.

Early in March I accepted Sutherland's offer of the Spring Valley post and sent off my letter of resignation to Hamilton. The latter was not pleased and wanted me to be off Oodoowerre before his return in mid April. On 24th March I wrote to my parents telling them that;-

"There are not many days left here now. I will soon have to take over Mai Mallay from 'Woodrow' Wilson, and the bungalow from Mrs 'Woodrow'. Yesterday Uncle Felix 'phoned me from Badulla where he had come to see the Government Agent. He wishes to come to stay the night on Sunday, and have a yarn. Very good of him to go to so much trouble. I told him I intend to try out a horse on Sunday. He thought it a splendid idea, especially if it will be cheap to buy in say six months time. A light car can get to within one and a half miles of the bungalow, and it is just possible to get a motor-bike all the way. Anyway a horse would be a much more healthy way of getting about the estate. As a matter of fact Wilson says he has estimated for a road to go past the bungalow this year. It would be for taking light Austin leaf lorries, but there are already wireshoots and an aerial tramway. A bike would be quite impossible As the hillsides are steeper than most house roofs.

"Mr. G-C came over yesterday and seemed to find everything to his satisfaction. He is very pleased with the new line-rooms and says they are the best on the estate, which isn't saying much! I have asked Mrs G-C if she will kindly come and look at the bungalow next week to see if it is properly aired and clean etc. She replied that she will be delighted to come; so is coming to tea. I have been painting wicker chairs in the evenings. The curtains, carpets and mosquito nets etc., have returned from the dhoby.

Yesterday I got an engineer to come and see some of the factory machinery which had been giving trouble. A roll-breaker needs new shafting and bearings and one of the sorting machines is laid up. The crop is still tremendous, and I now estimate a crop of 103,000lbs. of made tea for the month. The previous highest monthly crop was 78,000 lbs. So it will easily be a record. The teas are not good though. I have to roll some of the leaf when it is nearly green, as in this weather the atmosphere is full of moisture, and with driers being used to fire the tea at the same time it is difficult to get a proper controlled wither."

(Actually the loft space was quite inadequate to spread this quantity of leaf so the floors of the lofts and verandahs had to be used).

"It has been very nice, during the past few months, getting Saturdays off for CPRC parades and cricket, even if it does mean having to work on Sundays. The coolies have turned out regularly every single day since the middle of February to help get the crop in, and the factory has been working twenty four hours a day, or day and night.

Saturday 27th March 1938. This morning I went in to Badulla early to meet Polson who had sent his horse in the day before to the club stables. It is a very nice looking thoroughbred English mare of 14 1/2 hands, aged 11, named Favour. I tried to mount her in the club compound, but first she tried to bite me then let out a colossal kick. She kept trying her tricks and I had to dodge in and leap aboard. Once on she was as quiet as a lamb. Polson says he thinks she was once badly treated and always tries to kick strangers at first. To begin with I trotted her along one end of the race course, then cantered her. She went beautifully until we came to a herd of cattle which caused her to buck and shy a bit. Then I had another fast canter,and a go at some jumps which she sailed over. When

I dismounted she was friendly and only put her ears back for a moment. I told Polson I will take her, and he says I can just look after her until he comes back from England, then if I want I can buy her. He suggests Rs-150 (about £12.), but as I get an allowance of Rs.50 a month for horsekeep, and should be able to save say Rs.8 (about 12 shillings), I ought to be able to make up her cost within two years.

When I got back here I found your very nice letter of 18th. Mum. It is very clever of you to see Major Sutherland in your imagination exactly as he is; with a beaky nose, sandy moustache turning grey, and a smile in the eye!.

(In the end Favour never came to Spring Valley. It turned out that Wilson wanted me to look after his horse Charlie. Charlie was already in the stables at May Mallay, and the horsekeeper there knew him well and was accustomed to looking after him, so it seemed to be a more sensible arrangement for me to take on Charlie. The only concession I made was to change his name to 'Chas', so as to avoid confusion, between rider and horse).

My letter continued:- "I expect Guy (my youngest brother and the eighth and last of the family), will be at home for the hols, so I enclose a new stamp showing Sigiriya rock. I am expecting Uncle Felix any moment so will carry on later.

Monday, I had a very nice letter from Auntie Kill (Hilda) today She sent me £.1-0-0, which was very naughty of her, I will get some mats for the bungalow. Polly Polson makes beautiful rugs and mats, so if I buy him the material he will make me some. He does it to amuse himself in the monsoon evenings.

Uncle arrived in the evening and said how delighted he and Aunt Dollie were to get my letter saying I will be going to Spring Valley. He asked me to get a chit from Mr. Grant-Cook, as tactfully as I can, saying he considers I have kept the place in good order, and also asked me to get the vet to look at the animals to say they are in good condition. He says I ought to be able to save about Rs.75 to 100 a month, and it is a very good thing to accumulate money in the way you suggested, but as the Ceylon banks only pay 1 to 1.5 per cent interest, he suggests I should buy a few gilt edged securities once a bit of money is accumulated. Aunt Dollie is in bed with 'flu' but wrote me a cheery letter with the envelope addressed to "Charles, Kindly Favoured by the 'Polar'"!

So ended my first ever paid job, and I left Oodoowerre for the May Mallay (trans = High Field, or Hill), and Nalla Mallay (trans = Good Field, or Hill), divisions of Spring Valley. My time in acting charge had gone successfully thanks to the Staff and Labour Force who had put up with my inexperience. Unfortunately there is very little that I can remember about individuals after an interval of fifty three years. I do recall a couple of incidents involving some of the labourers or coolies as we used to call them. This term implied no disrespect any more than the old term 'postman' did for what is now called a Postal Delivery and Collecting Officer!.

One day whilst on my way back to the bungalow I came upon a man thrashing a woman in the tea. The woman was shrieking loudly as the man beat her with a stick. I shouted at him to stop, whereupon he started to abuse me and came towards me in a threatening manner. When he was near I punched him on the jaw and he fell as if poleaxed. The woman immediately started to shriek at me. After I managed to calm her I picked up the man who appeared only dazed. It turned out that they were husband and wife. From that incident I learned; (a) never to interfere between a husband and his wife, and (b) never strike any human being unless one's own, or some other person's, life or safety is in immediate danger. Nor did I ever again strike any of my workers. I expected repercussions from the labour force but there were none. In fact they may have been pleased as the man was known as a notorious bully.

The second incident occurred when I was in the field with a gang of young female pluckers, with the usual male kanganies (foremen). The young and pretty girls usually formed what was known as the "tippers" or young field pluckers. The 'young' in the latter instance referring to the age of the field from its last prune. The 'young' tea needed more careful plucking and the 'Kuttis' (girls), with their nimble young fingers were the most skilfull for this work. Soon after arriving in this field one of the kanganies told me they had come across some wild boar in the tea. I then called out, so that all could hear in my best Tamil (or so I thought), " Where are the pigs". There was a roar of laughter from the men and the girls all tittered and covered their eyes and mouths. I was at a loss to know what had happened, and only later learned that the name for a female private part is very nearly the same as for a pig,. To get the correct pronounciation the emphasis altered on one vowel. I had asked "Where are the female private parts". They must have thought me very ignorant and inexperienced.

Wanderoo or Ceylon Grey Langur.

CHAPTER FIVE

MALARIA

Any young person arriving in Ceylon to take up an appointment in the old days could not avoid becoming involved with malaria. He might be fortunate in avoiding infection, but he would see its effects everywhere, and would soon be recruited into some form of anti-malarial operations. A young planter would lead an isolated life and have to cope with all kinds of emergencies and situations, and would be dependent on his Indian or Sinhalese subordinates for much advice. He shouldered immediate responsibility and picked up the job as he went along. The nearest European was usually his 'Peria durai' or P.D, his immediate superior, so it was difficult not to follow his example which, fortunately, was generally good.

The year of my arrival in the island, 1935, saw over over five thousand deaths in our local town of Badulla alone, and that years' outbreak saw over 100,000 deaths throughout the island, owing to the reign of the King of Tropical Diseases in the form the Anopheline Mosquito.

Malaria is no longer a serious danger to whole populations in the tropics, and young people today have very little concept of its appalling effects on humanity right up to the late 1940's. So a short laymans' description of causes, symptoms, and measures taken on estates to combat this dreaded disease, and its earlier effect on the island's ancient civilisation, seems to be appropriate here.

Malaria is caused by the invasion of the blood by minute organisms called malarial parasites. These parasites, ingested with the blood of an infected person by the female anopheline mosquito, undergo a process of development in the body of the insect and, when transmitted by the insect biting a new host causes malaria in that new host.

Malaria was almost universal wherever warmth and water existed together and the anopheles would breed. There are three types of the disease, Tertian, Quartan, and Subtertian (malignant) malaria. The last is by far the most widespread form in the tropics, and is often fatal. The anopheles breed in sluggish streams, pools, empty coconut shells, tins, or anywhere else where water lies. Their activity is confined to dusk and the dark hours.

The symptoms of Tertian malaria, which has an incubation period of 5 to 20 days, consists of paroxysms of fever every 48 hours (i,e every third day), These have a cold stage,a hot stage and a sweating stage. During the cold stage the patient suffers from headache, nausea and yawning, which usually comes on early in the afternoon. The skin is blue and cold, pulse rapid, and there may be vomiting. This is followed by a severe rigor, and the patients temperature shoots up. The patient becomes flushed and complains of severe headache and thirst. The pulse is full and respirations rapid. This stage lasts for up to six hours, and towards the end the temperature starts to fall. It is followed by the sweating stage when, after profuse perspiration, the patient feels more comfortable and often falls asleep. The whole attack lasts from 10 to 12 hours, and the next paroxysm follows it at the same time of the day 48 hours later. After several of these attacks the patient becomes profoundly anaemic, and the spleen becomes enlarged. Sometimes there is double infection, Caused by two different mosquitos, so that the paroxysms are daily events. Again there may be a mixed infection of tertian (benign), and malignant parasites.

Quartan malaria is not nearly so widespread, nor is it so prone to relapses. It behaves much like Tertian, but one serious complication of this form is chronic disease of the kidneys.

Subtertian (malignant) malaria, which has an incubation period of 2 to 14 days, is the most serious form of the disease. It begins with symptoms somewhat similar to those of benign tertian, but the temperature tends to become continuous and irregular. After about three weeks of this fever, anaemia and debility are pronounced, and may become serious. The patient often has a tinge of jaundice, and may be delirious at times. Relapses tend to occur and, during any of these the patient may suddenly develop hyperpyrexia and become wildly delirious, or develop conditions closely resembling cholera, or he may become comotose and die before there is time for adequate treatment.

Quinine was the stock form of treatment for all three forms of malaria, but it was essential to adopt a strict bed routine, and for paroxysms to be dealt with appropriately, for each of the cold, hot, and sweating stages. A patient would often appear to have recovered completely, from malignant malaria, on return to a cold country; but if he should contract a chill, or become run down for some reason, the disease could flare up again after a long latent period.

This brief description serves to indicate that prevention of the spread of malaria was vital to the health and efficiency of any tropical work force, such as engaged in tea and rubber estates in Ceylon.

The early development of estates had been largely confined to the highlands, which enjoyed a remarkable freedom from the disease. The public health of these estates had been constantly threatened by cholera, plague, and small-pox, brought from their native land by the South Indian workers; but outbreaks of these conditions were first restricted, then prevented, by quarantine. Rubber was introduced into the island at the end of the last century, and was soon followed by the opening of estates in the mid and low country areas, which were not immune to malaria.

In the wetter areas, on the South West side of the island, the incidence of malaria tended to decline as the crops matured and provided heavy shade. But in the areas of moderate rainfall it became a seasonally recurring pestilence, whilst in the drier regions it formed a continuous deadly plague.

The serious wastage by malaria caused planters, in 1925, to organise, through the Ceylon Association in London, a visit to the island by the famous Sir Ronald Ross, to make a survey on the spot and suggest remedial action. This gave birth to a Malaria Control Scheme, and although the problem was confined to the mid - and Low - country, many Up-Country (highland) estates gave financial support "pro bono Publico", such was the concern amongst planters and proprietors. There followed periods of success, then set-back, hope, then despair. Oiling of the main rivers was the first remedy tried. Surveys were carried out on all malarious estates. Observation stations were established, statistics collected, and a service of advice and recommendations introduced. It was established that anopheles breed in pools during the dry season, that any extension of such weather caused a marked increase in breeding, and that plentiful rains washed the larvae away and prevented breeding until the next dry season.

However during the severe trade slump in the early 1930's there was a tendency to cease expensive oiling, and to rely on small doses of plasmoquine to keep the labourers healthy. Then after many years of regular rainfall the 1934 S.W. monsoon failed, and was succeeded by a N.E. Monsoon markedly deficient in rain. Thus the anophelese mosquito went on breeding unrestricted. Full scale antimalarial measures were resumed and with the plentiful rains of 1935, the epidemic gradually subsided, but not before causing the horrifying casualties mentioned earlier.

Dr. G.Macdonald of the Ross Institute, who had worked strenuously on the problem, left the island in about 1938 to become the first lecturer, then Professor, of the London School of Hygiene and Tropical Medicine. Fortunately, as Director of the Ross Institute, he was still adviser to the Ceylon Estates Malaria Control Scheme. His place in Ceylon was taken by Dr. R. Svensson DSO., MC., who continued the work so that, eventually, each estate was provided with a map showing potential breeding places, and details of the best methods of dealing with these danger spots.

Another widespread drought occured in 1939. Once again the incidence of malaria increased alarmingly, due at least partially, to a belated start in oiling by Government which had now undertaken to control all streams of ten feet width or more. The 1940 drought was even more prolonged. However oiling began in the preceding December and, for the first time, success was achieved in preventing anopheles breeding over wide areas.

The coming of D.D.T. after the second World War, and Lt. Col. O. J. S.Macdonald's advice for estates to carry out residual spraying (abandoning anti-larval measures), of human habitations and animal byres, with a 5% solution of DDT in kerosine, resulted in a dramatic and speedy fall in the incidence of malaria. The Government adopted a similar policy using the same preparation, This was very important as it cleared the villages near estates of the pest which would otherwise have continued crossing the boundaries.

With the virtual extinction of malaria the Planters Association was able to channel its health work into new areas such as Anaemia, the prevelance of Bowel Diseases, Child Care, and other health concerns.

Although the history of the struggle against malaria was unknown to me when I arrived in the island and went to Yapame estate as a "creeper', I quickly became involved in the daily routine of prevention, as the estate was situated right in a malarial zone.
The ideal method of prevention was to make breeding impossible by killing mosquitos in the larval stage. The anopheles larvae are short wriggling thread-like creatures which live under water but come to the surface periodically to breathe. Ponds can be drained and wells covered to prevent their

breeding, but pools in streams and elsewhere cannot be treated that way. However if oil is sprayed to form a thin film on the surface of the water the larvae will be unable to breathe and will die. Therefore oiling was adopted at regular intervals. Regular oiling gangs worked all over the estate, spraying oil onto pools, streams, puddles and ponds, and into water lying in drains. Coconut shells and empty tins were overturned. Estate workers quarters were visited to find out if any sickness was neglected or unreported. A sick "check-roll" was kept in which every dose of quinine administered was shown against every adult's or child's name. It was necessary to make sure the dose was correctly taken by every patient so the individual issuing quinine was ordered that he must put the quinine into the patients' mouth, and watch him swallow it. On no account was medicine to be handed to someone to take and give to a relative. Every individual had the medicine personally administered at regular intervals.

All this required a good deal of organisation and supervision and was, of course, additional to the normal routine work, but it was essential to prevent the spread of malaria and to treat sufferers. There was a dispensary and an apothecary on this small estate of 624 acres, The breakdown was 305 acres of tea, 60 of rubber, 75 of timber and the remaining 184 acres consisted of jungle and patna (grass land), The estate population would have been in the region of 700 souls. Serious cases were despatched to the nearest Government hospital. I was fortunate to escape malaria during my stay of nearly two years on Yapame. At Oodoowerre, my first paid billet, I was on the upper division, above the malaria belt. On Spring Valley, prior to the war, the May Mallay and Nulla Mallay divisions, of which I was in charge, were also in the malaria free highlands. My bungalow was situated at 4,800 feet above sea level. However, whenever we paid evening visits to Badulla, or to any neighbour living at low elevations, we wore mosquito boots, thick trousers, and long-sleeved shirts, and applied mosquito repellant to our hands and faces. Eventually I contracted malaria in India, during the war, but by the time Rosemary came out to Ceylon in 1949, the risk had become minimal, thanks to the advances of medical science.

Whilst estate workers and the inhabitants of the towns, and the more accessible villages, received expert treatment, the inhabitants in remote areas and in the jungles, were not so fortunate. The effects in the untreated areas were plain to see by anyone who went into the jungle or visited out of the way parts, especially towards the East Coast, called the Dry Zone. Leonard Woolf in his book "The Village in the Jungle", described the situation of helpless despair. "There was not a family in Beddagama which did not suffer, nor a house in which death did not take the old or children. The doctor Mahatmaya appeared in the village, bringing medicines, but his efforts were no more or no less successful than those of the village vederala. When at last the sickness passed away, it was found that the village had lost sixteen out of its forty-one inhabitants. The jungle pressed in and claimed two out of the eight houses, after the fever had taken the men, women, and their children, who lived there." Woolf goes on to describe the misfortunes of the remaining villagers until all had succumbed and the jungle encroached upon the last little hut, whose rotting walls and broken, tattered roof went down before it. "It closed, with its shrubs, and trees, and bushes, with the impenetrable disorder of its thorns and its creepers, over the rice fields and the tank. Only a little hollow in the ground where the trees stood in water when the rain fell, and a long little mound which the rains washed out and the elephants trampled down, marked the place where before had lain the tank and its land". The village was forgotten. It disappeared into the jungle, from whence it had sprung.

It is a fact that the demise of the ancient Sinhalese civilisation, which lasted two thousand years, was ultimately brought about by malaria, causing the jungle to envelop the great cities, and the islands' population to be decimated. The cities were lost to the world until the jungles were penetrated again by the British. Even today there are many fine ruins scattered in the jungles. Rosemary and I have come across such ruins on our jungle trips. Over the past century or so, the Ordinance Survey

Department has rendered great service in discovering, and recording, the sites of such ruins, and marking them on the maps. But in general the majority remain unknown to the public, and unexplored. In the 1960's , on a jungle trip, we came upon the Maligawila ruins which included a fallen statue of Buddha, 50 feet long, a stone bath, stone enclosure, and a Pokuna (a stone bath usually for religious observancies), with a flight of stone steps and a terrace. I believe this has been opened up since. It is about 10 miles South east of the small town of Buttala.

The vigour and fame of the greatest of the cities, Anuradhapura, was known far and wide. Ptolemy wrote of it, and drew a map of the country. Ovid mentions it as the last outpost of the world. Pliny, writing of Ceylon and its people said, "they possess a greater degree of civil liberty, and a greater regard for popular rights, than perhaps ever existed in any region of the East".

The fabulous historic riches of this great civilisation were all carefully recorded in an historical record called the MAHAVANSA, or Great Dynasty, covering no less than twenty-three centuries from the year 543 B.C. to A.D. 1758. It was the Hon. George Turnour, Government Agent at Ratnapura, in Sabaragamuwa Province, who in 1826 discovered the whereabouts of the Mahavansa, and the key to its Pali text, helped by a buddhist monk named Galle. He translated thirty of the one hundred volumes before his death, and opened the way for the translation of the remainder, and the discovery of the two most remarkable lost cities in the world, Anuradhapura and Pollunaruwa, as well as the translation of carved inscriptions found all over India. The books of the Mahavansa gave detailed cover of Ceylon history over the period during which fifty-four Kings of the Great Dynasty reigned over Lanka.

The records show that, among many other marvels, the agricultural progress for hundreds of years was phenonemal. The whole surface of the plains was altered by tireless industry, from wild and forbidding forest to an expanse of cultivated fields, roads, villages and towns. This was achieved by an immensely skilfull use of irrigation, of which Tennant said "the number of these stupendous works almost exceeds credibility." Throughout the annual drought, which in some parts of the north is of almost twelve months duration, the dispersal of adequate water continued without let or hindrance. The Sinhalese, busy with their vast agricultural schemes, and as Buddhists formally enjoined from killing, employed hardy Tamil mercenaries from South India to carry out their policing and military duties. It was not very long before the Tamils realised their strength, and as early as 237 B.C. two of their young men murdered the reigning king Suratissa, and usurped his power. From this time onwards internecine wars dragged on. Sometimes the Sinhalese gained the upper hand and, for various periods peace would be restored. But sooner or later other Tamil invasions took place. The irrigation works gradually fell into decay. The Tamils spread themselves all over the fertile plains, but their abilities were not equal to the task of maintaining a civilisation. so far above their own culture. The jungle advanced, and the previously controlled tanks (reservoirs) became stagnant marshes. King Mosquito began his merciless reign. Eventually the remaining Sinhalese fled to the mountains and died like flies as they went, so that when Dr. John Davy published his "Account of The Interior of Ceylon" in 1821, he recorded that "the population of the whole island does not exceed 800,000 souls".

When I went out to Ceylon the population was around four and a half million. In 1971, just after we left thirty six years later the census recorded over twelve and three quarter million people as Inhabiting the island. Now it must be at least fourteen million, but after the century and a half of settled rule under the British, internecine strife has gradually returned, with the Tamils of the North and East at bloody war with the Sinhalese and Moors. Who knows what pestilence or disease may reappear and take its toll once more?

After an attack one planter described malaria;-

"It comes creeping up a fellow's back like a ton of wild cats, goes creeping through his joints like iron spikes, and is followed by a fever which prevents the patient thinking of anything save Greenland's Icy Mountains. It isn't the now-and-again kind, but gets up with a fellow at daylight and sleeps in the small of his back at night. His teeth feel about six inches long, his joints wobble like loose wagon wheels and the shakes are so steady that he cannot hold a conversation except by putting in dashes. Then perhaps he gets better and goes on making his fortune; then gets worse and goes on digging his grave ...."

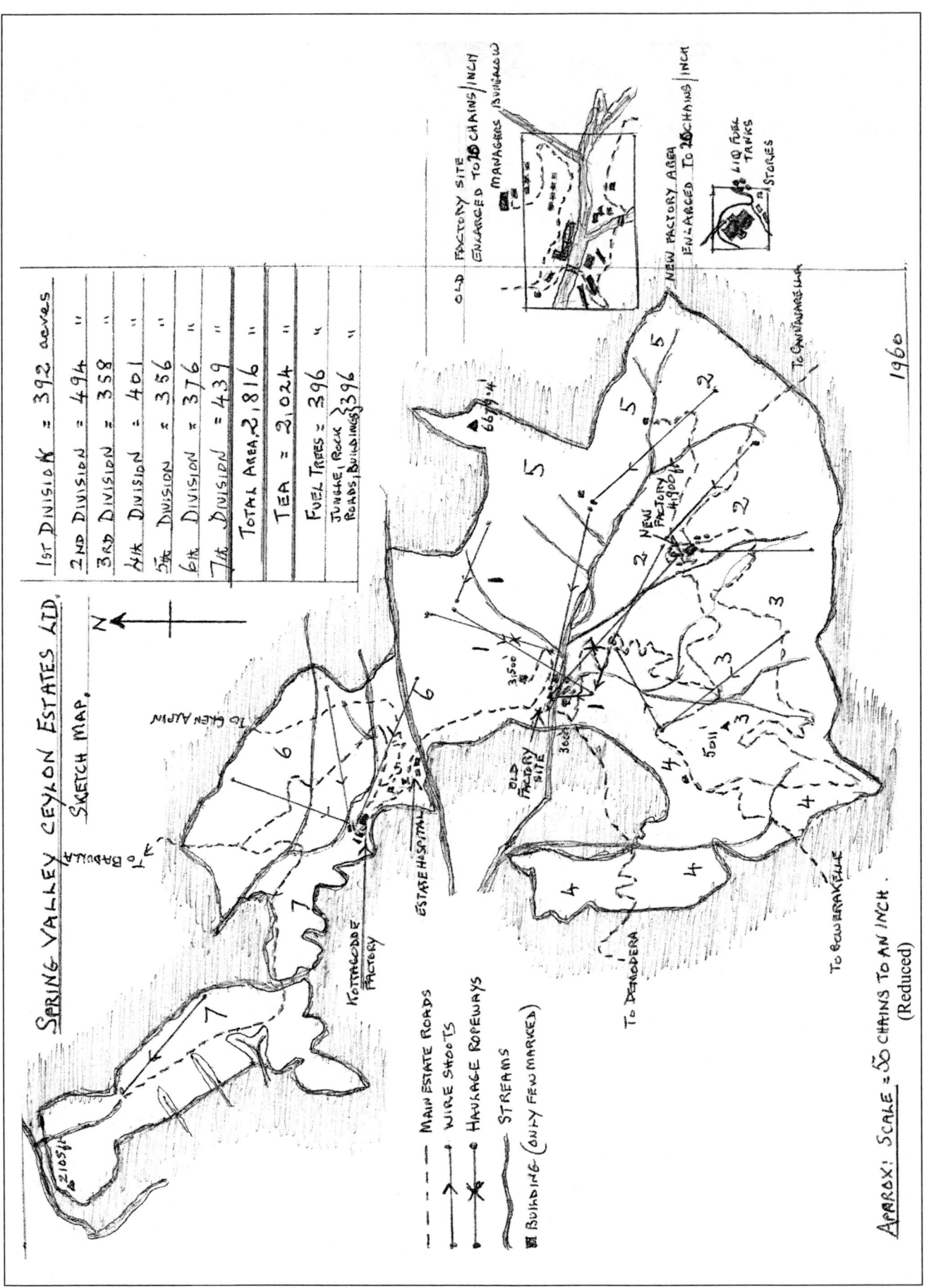

SPRING VALLEY CEYLON ESTATES LTD.

Sketch Map.

N

1st Division	= 392 acres
2nd Division	= 494 "
3rd Division	= 358 "
4th Division	= 401 "
5th Division	= 356 "
6th Division	= 376 "
7th Division	= 439 "
Total Area,	2,816 "
Tea	= 2,024 "
Fuel Trees	= 396 "
Jungle, Rock Roads, Buildings	} 396 "

Old Factory Site
Enlarged to 10 chains/inch

Managers Bungalow

New Factory Area
Enlarged to 20 chains/inch

Liq. Fuel Tanks
Stores

To Girandurakotte

1960

- - - - Main Estate Roads
↗ Wire Shoots
✳ Haulage Ropeways
〜 Streams
▨ Building (only few marked)

To Badulla

To Glen Alpin

Kottagodde Factory

Estate Hospital

Old Factory Site

New Factory H.4000 ft

66799 ft

To Demodera

To Ecumbrakelle

Approx: Scale = 50 chains to an inch.
(Reduced)

CHAPTER SIX

MAY MALLAY

"Tea is a work of art and needs a master hand to bring out its noblest qualities."

Okakura-Kakuzo.

The May Mallay bungalow on Spring Valley was built of cut stone with a red corrugated iron roof. Jock Sutherland had built it in the nineteen twenties, after demolishing the old wooden building on the same site. He quarried the stone from the rocky slopes behind, and had the timbers sawn from trees growing in the surrounding tea fields. Jock used to tell the story of how he arrived on Spring Valley in 1919; after returning from four years in the trenches, in France, with the Argyll and Sutherland Highlanders; and was sent by Wilfred Rettie - the Manager, who had also just returned after being demobilised -, to take charge of these top divisions, and to get rid of the old reprobate who had "held the fort," there during the war. Jock arrived up at the bungalow, after the long climb on foot, and knocked loudly on the front door. After some time, as there was no response, he went round to the servants quarters, at the back, and roused a sleepy cook appu. Jock told him to tell his durai that he had come to take charge, whereupon the old servant went round to the front of the building, which was built of timber and was raised about three feet above ground level on stilts. He peered under the floor calling out for master to come out. Eventually the man was found fast asleep, and snoring, underneath the centre of the bungalow. He was as drunk as a lord.

Both Jock and Wilfred had miraculously survived the bloody years in the trenches and both had been decorated for gallantry, with the Military Cross. But they did not get on with each other on Spring Valley. In the mid twenties Jock was given charge of Ellamalle Estate, an isolated property at the end of Matturata District. He remained there until his appointment to relieve Wilfred in 1931. Wilfred then became Visiting Agent and Agricultural Adviser. This made life very hard for Jock, who was determined to improve Spring Valley by implementing his own policies. Looking back on the V.As inspections, when I accompanied Jock and Wilfred round my divisions twice each year, I am surprised that the friction between these two passed over me almost unnoticed. It was only in later years, when I had access to Wilfred's reports, and Jocks comments on them to the Board of Directors,in London, that I learned the extent of the controversy. Wilfred had become a full-time Visiting Agent, with his headquarters at the Hill Club, in Nuwara Eliya. Not being personally involved with running an estate he rather lost touch with some of the practical problems. Nevertheless he was extremely astute, and knew everything that there was to know about tea, and Ceylon. He was born on Spring Valley, and took over from his father in 1913.

To get back to May Mallay bungalow; The front door opened onto a central passage. To the right was a fair sized sitting room with a bow window and a fire place. On the opposite side of the passage was a dining room. Both rooms had large windows opening onto the garden. These looked out towards the distant mountains of the Udupusselawa range, and down to the valley with the town of Badulla, far, far below. Behind these rooms, also on both sides of the central passage, were two bedrooms. Doors in. the rear walls of each of these led into bath-rooms, with further doors into lavatories. All the floors were of cement, as were the two large baths. The lavatories were furnished with "thunder boxes". The buckets of the thunder boxes were daily lined by the sweeper with aromatic foliage from macrocapa trees growing nearby. There were trap doors in the walls behind the thunder boxes, for easy access and removal of the buckets. Lighting was by oil lamps and candles.

At the far end of the passage, doors opened onto the back verandah. On the left of this back verandah were five or six steps up to the kitchen, where cooking was done on a wood fired "Dover" stove. Beyond this there were two rooms for servants. On the right side of the back verandah was the office presided over by the clerk. Here were two telephones. One crank-handled machine for internal communications to the muster grounds, and the kanakapillais quarters, the other connected to the outside world through the Spring Valley exchange. Further back against the hillside, beyond a vegetable garden, and at a higher level, were the stables. There were two horse-boxes, and a room for tack and feed, with space for the horsekeeper and a verandah along the front of the whole building.

The bungalow was furnished with bare necessities such as beds, tables and chairs. The Wilsons very kindly left some of their furniture and curtains for me to look after until they returned from Home Leave. This helped to cover the cold bareness of the place, and gave me time to collect some things of my own.

As the whole place was situated at 4,800 feet above sea level on an exposed ridge, it could get very cold, especially during the North East Monsoon. At this season the whole estate was often enveloped in thick mist for weeks on end. The spur continued upwards behind the bungalow to a ridge 45O feet above. This ridge separated the Kumbalwela Korale, to the East, from the Rilpola Korale to the West, the house being inside the latter Korale (District once under the authority of a Korala). The ground in front fell away steeply to the West for over 2,000 ft. Half a mile to the North were the almost precipitous slopes of Namunukula (6,700 ft), and close to the South lay the slightly less steep feature known as Kardumallai (5,650 ft). There was a constant sound of rushing water; for all around were numerous streams leaping down the mountain-sides in countless water-falls. Lower down these swelled the embryo Beddegama Oya which joined forces with the Badulla Oya near the bottom of Spring Valley, and then swept on to the Lugal Oya, and eventually to the great Mahaweli Ganga which makes its way to the sea just South of Trincomalee. During and after heavy rains the sound of the waterfalls became a mighty roar.

The annual average rainfall at May Mallay was 147 inches, the greater part falling between mid October and early April. The lower divisions of the estate received only 70 inches a year. Even this is more than three times the rainfall we experience in this corner of England.

I had been mistaken in thinking there was no motorable road as far as May Mallay. I found, on arrival there, that the road had just reached the bungalow, Though it was not quite finished it was just passable for small cars and vans. The drive up from First Division was awesome and I nearly scared the life out of many passengers during the following years. I was sometimes pretty scared myself when driving a motor-cycle late at night in thick mist and heavy rain. I have actually run into buffaloes, which are exactly the colour of the road, and knocked myself and my machine about sufficiently to have to limp the final mile to the bungalow.

In the days before motor transport, May Mallay had been used as a hot weather station by past managers and their families. They had planted many varieties of trees round the bungalow, and had established productive flower and vegetable gardens. It was the custom then to provide the other bungalows on Spring Valley with fruit and vegetables from May Mallay. In return I received free milk and eggs from the Sutherlands dairy, and pineapples, plantains, and other tropical fruit from the bungalows at lower elevations. There was a free allowance of two gardeners. I have never since seen such crops of strawberries and tangerines as there were in those pre-war years. I had strawberries for breakfast, lunch, tea and dinner. In fact I got heartily sick of them. After the war and the demise of animal transport, and animal manure, the productivity declined. So did the practice of exchanging produce. The garden also produced plums, figs, peaches, loquots, guavas, passion fruit, tree tomatoes, mulberries, lemons, oranges and grapefruit. Almost any kind of English vegetable would flourish. In a letter Home I described the garden as being full of cannas, phlox, antirrinhum, gladioii, pinks and carnations, besides bushy perennials such as hibiscus, poinsettia and bougainvillea.

I spent the first few days with the Sutherlands, as Jock wanted to show me the whole property, and to introduce me to the other planters before moving up to take over Second and Fifth Divisions (May Mallay and Nulla Mallay) from Wilson.

The "Big Bungalow" was reached, from May Mallay, by descending 1,800 feet to the main factory and then climbing up 600 feet again. I often used to get there by running down the ridge. In those days I could run all the way up again too. After the war, and service in a Gurkha regiment, I came back even fitter and could emulate those wonderful little men who virtually fall down the hills, leaping from rock to rock.

The Sutherlands kept a good table and I much enjoyed my few days with them. There was always porridge for breakfast and oatcakes for tea, besides the usual bacon and eggs and marmalade at the former, and thinly sliced bread and butter and cake - all home baked as the nearest bakers shop was many many miles away - at the latter. Jack Cranfield the tall and lugubrious assistant manager, showed me round Kottagodde and Nilgodde (6th and 7th Divisions), with its own factory and a 42 bed hospital which served the whole Group. Fields in the lower part of Kottagodde descended to almost 2,000 feet above sea level.

The steep strips of jungle which protected the streams running through these lower tea fields were rich in tropical bird life. This gave much joy to Jack who visibly brightened whenever he walked through this section. Normally he was extremely pessimistic by nature and, as a result, was known as "Happy Jack" by his contemporaries. He was called "Mooken Durai" by the labour, meaning the master with the big nose. Jack also showed me round the main factory, at 1st. Division. Close to this were the blacksmiths forge, carpenters workshops, and the lorry repair garages. The main rice store, the main diesel fuel storage tanks, petrol pumps, and estate shops were also situated here. The main estate school was several hundred feet above the factory.

Gerald Du Pre Moore (Dopey) was in charge of the field work on this very steep division. His transport was a 35Occ B.S.A. motorcycle, but the paths on the "North Face of The Eiger" as it was known, were impassable even for Dopey and his machine, so Shank's Pony was the order of the day. You could actually get near his top fields by ascending some 1,500 steps, but most preferred the zig-zag paths.

Ian Fletcher was away on a jungle shooting trip so I didn't meet him, or visit his Nawalawatte and Pudumallay Divisions (3rd & 4th), until later. Jack Cranfield's wife, May, was the daughter of Missionaries in South India. Both she and Myrtle Sutherland were delightful people who did everything possible to help me settle in and feel at home. Jock would have been about 45 years old, Jack around 40, Ian Fletcher perhaps 30, and Dopey was my senior by just a year. Jack and May had had a terrible time on an isolated rubber estate in Travencore before they came to Ceylon. The company went bankrupt and there was no money to meet staff and labourers wages, or even to provide basic food. Jack stood by his labour force, refusing to pack up and leave until some sort of provision had been made. He was a sick man when he arrived on Spring Valley, and May was none too fit either. It was probably that experience which set his pessimistic mood.

After this I spent a day or two going round with Wilson. Then I was on my own. The china and glass, bedding and other equipment I had brought out from home so long ago were all unpacked at last. I received an unexpected visit from Mrs. Hamilton, who came all the way from Oodoowerre expressly to find out how I was coping. Seeing my lack of cutlery and cooking utensils she insisted on sending some of her own over by her driver, Joseph. The cutlery and some pyrex ware, which she declared to be surplus from Oodoowerre along with some cruets, are still in daily use in our cottage kitchen in Suffolk in 1991. I soon became very attached to that home. My feelings were exactly expressed by another planter, in Assam, who long before had written;-

"I loved that bungalow more than any of the other habitations I have infested before or since. It fitted me exactly, and above all it spelt freedom with a capital F..... With a good dog and no lack of books from home the talk of loneliness out in the jungle being bad for a man and sending him slightly crackers is absolute nonesense .... I had more than enough to keep my mind occupied"

I acquired a dog very soon. It was a wire haired fox-terrier from a litter belonging to the Provincial Survey Officer in Badulla, a Mr. Tippets, I think his name was. Pip was the most intelligent of the many dogs I had. He was very sporty, and as brave as a lion. Once, when chasing a monkey he went clean over the edge of a 100ft. precipice. His fall must have been broken by tree tops and bushes below because he reappeared, not long after, showing little damage from his fall. He loved to jump up and lick Chas's face when I was mounting, and he often rode on my motorcycle, sitting on the petrol tank in front of me. After a few months I acquired a second dog, a small daschund which I named Adolf. Gerald ru Pre Moore had quite a menagerie. There was his smooth haired fox terrier, two Daschunds, a mongoose which was suckled by one of the daschs when it was found in the tea as a baby. He also had a palm civet cat and a mouse-deer (tragulus meminna). When Jock and Myrtle went on Home Leave the following year I took care of their dour old Aberdeen and Cairn terriers.

In due course the Wilsons returned from furlough and went to Ellawatte Estate, near Ella. I had to send Chas over to him and set about finding a new mount for myself.

The author with "Trickster" and terrier "Pip" at Nawalawatte, 1938.

May Mally bungalow, Namunukula covered in cloud behind, 1938.

Stables at Oodoowerre.

Major (later Lt. Colonel) F.I.S. Sutherland OBE, MC, ED, and Hugh Breay, near Ramboda on the way to a C.P.R.C. parade at Kandy, Spring 1939.

After advertising in The Times of Ceylon I eventually decided to buy Trickster an ex Australian race horse of 16.5 hands. Trickster was being looked after by Mansergh Hodgson, the man I had met when he was out of work at Oodoowerre. He was now on an estate in the Dolasbage District, on the Western side of the island. I agreed to pay Rs.150 and,have the horse for a month or so, and if he suited me I would keep him. I sent my horsekeeper by bus, and arranged for stabling, forage and water, at various points on the route across the island. It took four days to walk the horse over to Spring Valley, via Hatton, The Horton Plains, Wilson's Bungalow and Welimade. Trickster arrived in good condition and was a useful estate hack until the end of 1941 when I left him in charge of Colonel "Johnie" Maxwell Johnstone. Trickster had to be destroyed in about 1944 when there was insufficient fodder for horses. It says a lot for the horsekeeper and the Rest House keepers who attended to Trickster on that journey, that they carried out instructions sent to them by letter. I doubt whether anything like that would have been successfully completed in the 1960s or later.

About once a week Jock would meet me on my boundary after an early breakfast, and we would ride together to inspect a selection of works and fields, and discuss some of the many problems which arose almost daily. Unlike most estates in Ceylon we had no Head Kanganies to control the labour. The Head Kangany system on Spring Valley had been broken many years before. Shortly before the turn of the century, A.T.Rettie, father of Wilfred, and manager for 35 years, settled the debts of every one of the work force, and broke the Head Kanganies power over them. In their place he appointed Sub Kanganies who were the heads of families. These sub kanganies had to look after and control all their relations, and see that they worked regularly. To encourage them in this duty a "pence money" allowance was made, of a few cents for every day a gang member turned out for work. To discourage indebtedness a regular Savings Scheme was set up. Every worker was encouraged to invest at least a rupee monthly in this Savings Bank. There was no upper limit and interest was paid at the rate of 4.5 per cent a year, a handsome rate for those days. The labourers quickly realised the benefits of the scheme. Everyone was issued with a pass-book with these books being called in annually for audit. By the time I arrived some men and women had quite substantial savings. Withdrawals for funerals,weddings, coming-of-age ceremonies and the like were allowed on a fixed scale. Pensioners who retired to Coast (India), would take back all their savings, and they some-times had enough to buy a bit of land for growing rice, with a house, and still have sufficient left over to live in rather better style than the average villager. To safeguard these people from robbers, con-men and the like during their journey to their Indian village, the Estate Management could transfer the funds to an agent over there, for drawing by the pensioners on arrival. The agent in India would also carry out enquiries about old employees on behalf of relatives or the company in Ceylon, so quite an elaborate communications system was established between the estates and the villages in South India. This was extremely useful when the time came to recruit fresh labour as they would come willingly and have confidence in their future, and would probably be coming to join relatives.

Of course we had no powers by which we could force workers to carry out orders, such as the Army Act, or King's Regulations, and during the 1930s there was very little legislation defining the rights of workers and powers of employers. It would be nonsense to suppose that hundreds of estate workers could be intimidated into doing hard work by any one man. The only way to get good results was through mutual understanding and sensible cooperation. When treated right there is no better worker than the Tamil from South India.

With a range of nearly three thousand feet from the bottom to the top of my divisions, and having regularly to cover the 600 acres of tea plus another 200 acres of timber and forest in this very rugged terrain, worked by at least 600 people all of whom were resident with their families, I had very little time to sit and twiddle my thumbs. Any idleness on my part would have been instantly noticed and advantage taken. Whenever a new planter arrived on an estate the workers would watch him carefully and whilst summing him up, would carry on working as if no change had taken place. Eventually, perhaps after several months, when they reckoned they had his measure, they would decide to test him.

There were many ways of doing this. Perhaps a man or woman would come up clutching his or her stomach saying "mitchtim Voutu Vully, Laisse Vailey koodanger" (I have very bad stomach-ache, please give me light work, sir). The obvious reply should be "Go and see the doctor, and he will treat you and, if necessary, will give you a chit advising light work". That would put an end to the matter. However had an inexperienced or soft answer been given, such as "Oh Dear! you had better go and do such-and-such light work" then there would have followed a regular inundation of similar requests by half the labour force. Another try-on might be a request from an individual to be allowed to keep a goat in his line garden. Everyone knew that to feed the goat the owner would have to cut fodder from estate shade trees growing in the tea, and that once one goat were allowed it would be almost impossible to resist many other requests for a similar privilege. once goats are allowed on an estate immense damage is caused. Goats will eat anything and everything which comes their way and have been the cause of many deserts.

Rice was issued weekly to the whole labour force and their families. Sometimes complaints of short measure would follow. My immediate response would be to enquire whether the complainants had been present in person to receive their rice. If they admitted having been present I would ask why they had not complained at that time. Rice issues need careful control and supervision. A dishonest rice-store-keeper can make a fortune, over the years, by cheating every-one of a fraction of his measure at every weekly rice issue. The amount of rice in the measure can be varied by pouring the rice into it very slowly, or from a hieght, and by the method of "cutting". That is by drawing the cutting stick across the top of the brass-bound measure. These "measures" too need regular checking to ensure that the brass binding has not been removed from the top; the measure filed down, and the brass replaced. Another way to cheat is to dampen the rice to make it swell. The ways of cheating are legion, but when, in later years, I visited an estate with a very fat rice-store-keeper I would be suspicious. The method of measuring rice by the bushel, and fractions of a bushel, went back into history. In about 1969 the government decreed that rice issues would in future be weighed. This reduced cheating considerably but did not eliminate it by any means.

Had this sort of malpractice been overlooked it would rapidly spread to other departments. Then life would become difficult for everyone, with many disgruntled workers bringing complaints. The cheated workers would resort to all kinds of tricks to get their own back, leading to endless trouble and strife and an unhappy labour force. Labour troubles usually seemed to occur when work was short; that is when they had time to put their heads together and scheme. Generally the harder they were worked the happier they seemed to be. I am describing the days before the advent of widespread trade union power.

As could be expected there were a great many agricultural tasks which had to be performed in order to keep the tea bushes in good health and flushing well. One of the most important was pruning. Only the fittest and best male workers were selected for this task for which they received a special rate of pay. The tea bush needs pruning at regular intervals if its yield of leaf is to be maintained on a paying basis. After it has been plucked continuously in seven to nine day rounds, over periods of from two to five years, depending on the elevation above sea level, the climatic conditions of the particular estate, and the "jat" (type) of the tea bushes, it loses vigour, and the growth of flush falls off. This is due to the formation by the bush of such a quantity of wood and foliage that the available supply of sap is insufficient to enable plentiful new growth. Too many "banjies"(blind shoots) form and not enough bud. If, however, the superfluous wood and foliage be removed by pruning, the bush will regain its vigour and proceed to put on a fresh head of young foliage.

To ensure that annual crops are produced more or less evenly it is advisable to arrange for the annual acreage under pruning to be as near the same as practicable. For this purpose a pruning programme is drawn up covering several years ahead.This can always be adjusted if special circumstances arise.

It was the subject of pruning which generated the greatest amount of controversy between Jock Sutherland and Wilfred Rettie. For years Wilfred had adopted a policy of light pruning. As a result the bushes throughout the property had reached an undesirable height, difficult to pluck. Diseases such as shot-hole-borer (Xyleborus fornicatus), canker, etc., had increased alarmingly. Jock was determined to rectify matters through a policy of severe pruning with a view to a long term increase in yield. He drew up a programme of gradual and systematic cutting back. Looking back, I think the severity of his prune was rather overdone. However, there is no doubt that with the hard pruning and generous applications of fertiliser the condition of the tea improved dramatically during his period of management, and for years after his departure on retirement. On the other hand the tea prices obtained during Jock's regime never matched those during the time of Wilfred. Rettie adopted the light prune in order to reduce crop under the internationally agreed crop restriction scheme. His idea was to keep maximum acres in bearing whilst reducing the crop to the alloted quota. When Jock took charge he saw that this was a golden opportunity to prepare the estate for the time when full production would be resumed.

With pruning gangs being made up of the fittest young bucks on the estate it was only natural that these gangs were the most high spirited, and likely to be a source of trouble if not carefully watched. I sometimes promoted a leading troublemaker to Kangany, on the theory that he was likely to be a potential leader who was frustrated at the lack of possibilities for advancement. Once promoted he was expected to lead his erstwhile troublesome pals, but now by encouraging them to conform rather than to rebel.

Percy Will, then the manager of the neighbouring Ouvah Company, was a tiny man. I remember once sharing a two berth sleeper on the night mail to Colombo, with him, and noticing that his shoes on the floor beside his bunk seemed to be less than half the length of my size twelves. An amusing story used to be told on May Mallay of how Percy got the better of a troublesome ringleader when he was in charge there. As this happened in the early nineteen twenties and the story was told to me by one of the Kanagnies in about 1938 it must have impressed the labour force considerably. In those days there was one huge pruner, of most unusual size and strength for a tamil labourer. This man, Vemban, was constantly causing trouble by encouraging the pruners to disobey orders, to down tools at awkward moments, and to do generally shoddy work etc.. He was the bane of Percy's life and because of his unusual size and bellicose character had considerable influence over his fellow workers. Eventually he went too far. One day he went up to Percy, in the pruning field, making threatening gestures with his pruning knife and loudly abusing him. They finished up facing each other on a field path. On one side of the path there was a steep drop. All the other pruners were watching and awaiting the outcome. However Percy, who had a fierce temper when roused, was not going to let Vemban get away with such insubordinate behaviour. He shifted into a position so that he had his back to the bank, and Vemban faced him with the outer edge of the path behind him. Percy roared furiously to Vemban to "get back to work", at the same time advancing towards him. Vemban, taken by surprise, retreated a step, to the very edge of the path. Percy then advanced another step, letting out a second yell which caused Vemban to step back into space. He disappeared from view and was heard to crash into the tea twenty feet or more below. At this Percy let out a roar of laughter and was joined by the whole gang of pruners,

who slapped their sides with merriment. Nothing more was heard of Vemban for many days. When he eventually returned he was a subdued man who caused no further trouble, and from that day Percy's pruners worked well and happily. Ernest Percy Will was a great character with a fund of amusing stories. After retiring from planting in about 1950 he joined our Board as a director and always took a special interest in the affairs of the planting and estate staff whose problems he understood so much better than the majority of board members.

Cultivation and manuring were obviously very important functions on tea estates. As soil conditions in the tropics differ considerably from the average cultivated soils of temperate climates a few words of explanation seem desirable. The most striking differences are the much smaller content of lime and phosphoric acid. Lime is the base which is most rapidly leached out of a soil, and Ceylon's heavy rainfall, and comparatively high temperature have the effect of depleting the soil of lime to a marked degree. This has resulted in a higher percentage of magnesia than lime in most Ceylon soils, a condition which is seldom met with in temperate climates. It should not, however, be assumed that because of this all Ceylon's soils need lime. A low lime content is the normal condition of Ceylon's soils - as it is of most tropical soils - and the vegetation native to the island thrives under this condition. Its requirements should not be judged by those of crops grown in temperate zones.

As regards other constituents, the same variations are found among Ceylon soils as among those of most other countries. For instance, Potash is high in those containing much potash bearing felspar or mica, and may be low where these minerals are absent. Similarly, Nitrogen is usually high in soils containing much organic matter, and low in the quartzy ridges where organic matter does not accumulate. On the whole the chemical composition of Ceylon soils is poor compared with those of higher latitudes, and it is more to the climate than to the soils that we owe the luxuriant growth which is associated with the tropics.

Of equal importance with the chemical composition of a soil is its physical condition. On this depends its power of retaining moisture and of absorbing fertilisers, and also its temperature and aeration. These last largely determine the nature and vigour of its bacterial life. A soil to be in good condition should have its surface loose and friable, so as to allow easy penetration of rain-water and air, and to prevent evaporation during dry weather. The sub-soil should be sufficiently compacted to promote the capillary rise of subsoil moisture to the vicinity of the roots. This state of affairs requires frequent and regular cultivation.

In order to assess the responses to manure applied on various areas of an estate a very detailed statistical analysis is required, taking into consideration, amongst other factors, the types of mixture (the balance or imbalance of N.P. and K.), the proportion of V.P. tea to seedling tea, the proportion of good and poor seedling tea, the levels of manure each of these categories have received and their respective yields, and most important, whether the tea is looking robust or tired. It is not sufficient to look at the figures, on paper, of the over-all level of Nitrogen per acre applied, or the over-all Nitrogen per 100 pounds of made tea, or the over-all yield. In other words the assessment can only be made by careful inspection of the growing tea and knowledge of the conditions under which it is grown. If these matters are dictated from an office in Colombo or London serious mistakes will be made and the condition of the tea is bound to suffer. All these things would take a long time for me to learn.

Owing to the exceptionally steep and rocky nature of most of the terrain in tea, as well as the necessary close planting of the bushes, the cultivation of the tea fields has to be undertaken by hand-forking. This is done across the contour of the land to assist penetration of rain-water and fertilisers. The forker puts his four pronged fork into the ground to the full eighteen inch length of the prongs, then pushes the fork forward to make an opening into which he pushes the previously laid fertiliser, the Toppings from the green manure trees, and any dead leaves and prunings. On several occasions in Colombo I have met farmers from Britain and America who have expressed disbelief or scorn when told that we used no mechanical cultivators. I have always invited them to visit Spring Valley to see for themselves the conditions under which the tea bush is grown. They have usually arrived with great expectations of achieving some sort of mechanisation but in every case have departed defeated. I have been told by planters from the Darjeeling area of India that none of their tea slopes compare with Spring Valley for rugged steepness. When John Ferguson described the visit to Spring Valley by the Governor, Sir Arthur Hamilton Gordon in 1886, he pictured the scene on this up-on-end estate, with a rise of over 3,000 feet from bottom to top. He noted the detail of the workers having to tie themselves to trees in some places where most people would prefer not to stand - let alone work - without holding on!

Finally it is necessary, in this brief note about tea's unique cultivation needs, to stress that all systems of manuring must be based on the fact the product to be cultivated is a permanent one. The crop is gathered all the year round, except for a few months every three or four years after pruning. Manuring programmes, in consequence, must be framed with a view to the lasting benefit of the bush, and not merely with regard to the maximum crop that can be forced from it temporarily.

Of course there are many more field works that have to be attended to and I will briefly name just a few. Weeding was generally carried out on a contract basis. This work afforded very useful additional income to families. At week-end, and after normal working hours whole families would turn out to weed a four or five acre section given out at a fixed monthly rate. In later years this work was taken from some families and chemicals applied instead. Many varieties of pests, blights and diseases attack the tea plant. These can be classified as leaf diseases, root diseases and insect pests. The majority of these flourish in the jungle from which they travel, or are carried by the wind. It was only in 1946 that the most serious threat to the health of the bushes reached Ceylon in the form of the fungus known as Blister Blight (exabasidium vexens). Further mention of attacks by this fungus, which had the makings of a major disaster, will be made later. The most prevalent stem diseases are die-back, branch canker and stump rot. They are usually caused by fungi that attack the wood of the bush at points where the bark has been injured, where a branch has been pruned off, or galleries pierced by shot-holeborer.

A number of root diseases attack the tea bush, of which the most common are Fomes Lomaoensis, Ustulina Zonata, Poria Hypolateritia, Rosellenia Arcuata and bunodes and Rhizoctonia bataticola. The origin of most of these is to be found in the decaying stump of an old jungle tree, grevillea, acacia etc., from which infection spreads to the roots of the tea bushes. There is little chance of recovery once the roots have been infected by the spores of these diseases. Where an attack occurs it is necessary to remove the infected bushes and all their roots and burn them "in situ". Adjacent bushes whose roots may have come in contact with the diseased bushes must also be uprooted and burnt with their roots. The incidence of infection can be greatly reduced if all trees growing in the tea are ring barked about eighteen months before removal. Properly ring barked trees die slowly and are unlikely to be host to the spores of poria, ustulina, rosellinea etc. To achieve success good organisation and close supervision are essential.

The main insect pests are Tea Tortrix (Homona caffearia), Shothole-borer (Xyleborus fornicatus), several varieties of Nettle Grub (e.g. Natada nararia), varieties of Termites (e.g.Calotermes militaris and different kinds of Mites. Suffice it to mention here that the methods of dealing with nearly all insect pests have greatly changed during the past forty years.

Very big changes have taken place in the propagation and planting of tea. Up until the late nineteen forties the tea plant was almost always propagated by means of seed. Originally the seed was obtained from N.E. India and to a smaller extent from China. Later many Ceylon estates started to grow their own seed and in 1923, with the entire prohibition of importation of seed from India, owing to the incidence of blister blight in some of the tea districts there, seed in Ceylon was in very short supply and fetched astronomical prices.

Nobody knew the extent of the life of the tea bush. At that time many in Ceylon were over 70 years old. Obviously they would not go on for ever so planters began to think about the possibility of uprooting some of the older tea and re-planting it with new material. Experimentation at vegetative propagation (V.P.) of tea started on Spring Valley in 1939. Peat Moss was specially imported from Siberia for this purpose. This V.P. work was carried out on a very small scale and was beginning to show signs of success when it all had to be abandoned owing to an acute shortage of supervisory staff. With the outbreak of hostilities many of the younger planters were involved with the defence forces or, if on leave joined up at Home. The remaining staff had their hands more than full at just keeping the estates in production.

On my 2nd and 5th Divisions a start had been made at mother bush selection. The aim was to find a "Golden Bush". This would have an exceptionally high rate of yield, its leaf would produce teas of high quality and flavour, it would be resistant to diseases and, very important in Uva, it would not suffer unduly from the effects of drought. The method was that, after the pruners had chosen their ideal bushes, the K.P. would inspect them and discard say 50% of those he considered least promising. I would then check those remaining and, perhaps, discard a further 50%. The remaining bushes would then be marked and code-named or numbered. Special pluckers would be allocated to pluck these potential mother bushes weekly. The product of each bush would be put into a bag and individually weighed before being separately manufactured on miniature machinery specially built for the purpose. Careful records would be maintained of every aspect of a bush's performance. The object was to assess each mother bush over a whole pruning cycle of perhaps four years. Any which did not reach a high standard would be discarded. War broke out just as these experiments were becoming extremely interesting. All experiments ceased and were not resumed until well into the nineteen fifties. Only in 1960 was it possible to open our first experimental replanted plots. By this time some of our tea was approaching its centenary and was showing no signs of flagging. In fact the bushes generally looked better than ever and yields had doubled during the past two decades.

As tea is seldom, if ever, nowadays propogated from seed, and seed-bearers no longer perform a useful function, it may be interesting to record, in outline, the practices involved before they are entirely forgotten. Tea planted for the production of seed should not be pruned, as required for the production of a leaf crop. Instead they must be allowed to grow into a tree 15 to 20 feet high, Depending on elevation and local climatic conditions they come into maturity in from 8 to 12 years. As a well grown tea seed bearer will cover a considerable area of ground they are usually planted at about 200 to an acre, or say at 15ft by 15 ft. This compares with 3,500 to 4,500 bushes per acre for normal tea.(Nowadays the figure is often much higher).

Although seed bearers may produce flowers throughout the year the blossom cannot set in very wet monsoon weather. When the capsules ripen they split and the seed falls to the ground. It is then collected by hand, usually by small 'podians' (boys),and taken to a central spot to be packed. Freshly gathered seed is liable to ferment if piled in heaps or packed when damp. All freshly gathered seed is tested by floating.

Floating seed is kept separately from those that sink. Some floaters do produce satisfactory plants but the proportion of failures is usually high, and if included in a consignment for sale complaints of low germination may be received. If seed is to be sent any great distance, or retained in store for a time, it should be carefully packed in dry charcoal in wooden cases (local Albizzia tea chests are suitable), and these need to be hermetically sealed to prevent premature germination. Germination is usually carried out under cover, after having spread the seed in a single layer on specially prepared beds topped with about two inches of fine sand. The seed is covered thinly with more sand and protected from sun and heavy rain by "cadjans" (palm fronds), or some other suitable covering spread on top of "pandals" (canopies). Moist conditions are essential for successful germination. Good "jat" seed should yield from 12,000 to 18,000 plants per "maund" (80 lbs), while inferior kinds may furnish much more, but weaker material. Depending on temperature and moisture in the beds germination may take anything from two weeks to a couple of months, and germinated seed needs to be picked over regularly. Any which has developed a spike should at once be planted in a basket, other container, a nursery bed, or as seed at stake. Nearly all seed bearers and seed nurseries used to be situated at the bottom of isolated valleys due, probably, to the fact that the only level land and constant water supplies were located there. This did not aid efficient supervision. The old seed nurseries on May Mallay were at the very bottom of the division on the site of the old cattle sheds which had been built in the days when all the estate's produce was taken to Colombo by bullock cart. I had the new V.P nurseries established close to the bungalow and near to the lorry road for ease of transport.

Preparations for planting out young tea seedlings and young V.P. clonal material in the field are basically similar. Holes 18 inches by 10 inches are dug in rows or across the contours. These are checked for size and depth before being carefully filled and marked with a stake until the seedlings, or young plants, are planted out during suitably wet or overcast weather. Prior to holing the area will have had-roads and drains cut and built. Terracing will be carried out and, if it is an area of old tea the whole section will have had its soil "rehabilitated" by mulching for about two years with suitable grasses or other leguminous crops. All young plants must be handled very carefully. In fact, the less they are handled the better. "Tay kannu sontha pillai patarama pudditchu mardray podu" (Treat the young tea plants as carefully as your own child) is the watchword. After planting every plant should be protected by stakes , baskets or ferns for shade.

In spite of their being seven lorries, a heavy and a light road roller, an ambulance,and various cars and motor cycles,as estate vehicles on Spring Valley at any one time,I do not remember a single serious accident during my 33 years on the estate. The mechanics kept everything in first class order and breakdowns were few and far between. A rigorous checking and servicing system was enforced. Every vehicle had a monthly service and a full overhaul annually with details carefully recorded on each vehicle as well as in a maintenance ledger which received regular inspection.

This accident free record says a lot for the mechanics workmanship since the very steep gradients and constant use of brakes and low gears caused well above average wear and tear. Mr. R.P. Martil, the Head Mechanic was the son of a South African War Boer prisoner. He successfully maintained some old 1919 Leyland lorries well into the 1940s, and saw that some of our 30cwt Morris lorries ran trouble free for 25 years. I bought a Wolsley Hornet Daytona Special sports car, second hand, soon after settling in, and Martil used to tune this for me in his spare time in return for a small payment or an occasional bottle of gin. He seemed to enjoy doing this and even made and fitted a radiator fan as the car used to boil excessively on the long steep climb up to my bungalow. I had to back to get round every one of the nine hairpin bends. These were widened later on so that you could just get round in one if you judged your approach right and had a reasonably good lock. By the way, that car cost me Rs.325 or about £26.

One of the original lorry drivers - after bullock carts were replaced - was V.Sinniah who caused much amusement among us planters. He was known as "Stand-up and-beg". Coming up or down a steep slope on a 9 feet wide road you would meet him on a sharp corner. Sinniah would take both hands off the steering wheel, stand up in the cab, and salaam, before resuming his seat and stopping the lorry. Fortunately both vehicles would seldom be moving at more than 10 m.p.h. Even so it was a miracle that he survived accident free. Personally I have always felt far safer on those mountain roads than on any motorways in Europe. A feeling which visitors never shared. Sinniah eventually retired and his lorry was taken over by his son S. Rasiah. Rasiah, like his father, was a good natured and simple fellow and a fair mechanic. We sometimes took him along on our jungle trips if we needed a back up vehicle. On one occasion,many years later in the 1960s, we were staying at the "Cuckoos Nest" a small cadjan thatched bungalow on the dunes near Arumagam Bay. We had with us our two children and three relations who had come down from Pakistan to stay with us for Christmas. While we were eating dinner the lights dimmed and we heard the ancient lighting engine slow almost to a stop. Suddenly it picked up speed again and the lights shone brightly, or rather as brightly as they ever did. This lasted for a few minutes only before they waned again almost to extinction. Again the engine picked up and with it the lights. This waxing and waning and change of running speed in the engine sound continued irregularly until I decided I ought to leave my dinner and see what was going on. To my surprise, on entering the engine shed I found Rasiah.bent over the engine covered in oil and perspiration. He was keeping the engine going by literally acting as a human tappet. He had his hands on the tappets and was manually helping them to move. It was a strange sight to see, in the dim light, his hands vibrating up and down at great speed and to watch the expression of deep concentration on his begrimed, perspiring face. When I suggested he ceased his efforts and that we get out the oil lamps he seemed quite disappointed.

Carupiah was a mechanic who, for some reason or other, had the privilege of being the sole driver of the six ton diesel road roller when road maintenance work was carried out, usually in the dry season. Carupiah did not possess a driving licence. This was of no special consequence since all the roads were the private property of the estate. Nevertheless he was very skilled at operating the roller on those fearsomely steep inclines. On one occasion whilst working on a section of road overhanging a precipitous slope the brakes failed and the roller took charge. Instead of saving himself by abandoning the great machine, for it to crash several hundred feet over the "khud" (gorge or ravine), he hung on to the controls. After careering round a couple of bends he came to a fork with one road going on downwards and the other sloping upwards. Carupiah managed to steer onto the up-hill branch, and so brought the Barford safely to a halt. This 1931 Barford Roller was saved from destruction thanks to Carupiah's bravery and devotion. Thanks also to his mechanical care it was still in regular use when I left the island in 1971 and, for all I know may still be in working order in 1991.

After.

Half completed.

Road and bridge building at Spring Valley

Jumbo makes less noise and mess than a bulldozer.

Before.

Rough country.

Small boys, delight in watching steam-rollers at work, but to watch another monster working was far more fascinating. I could spend hours watching elephants working. The late Field Marshal Lord Slim made no exaggeration when he wrote, in a Foreword to "Elephant Bill" the book by the famous elephant expert J.H. Williams :-

"It is not, I think, a matter of size and strength. It was the elephant's dignity and intelligence that gained our real respect. To watch an elephant building a bridge, to see the skill with which the great beast lifted huge logs, and the accuracy with which they were coaxed into position, was to realise that the trained elephant was no mere transport animal but indeed a skilled sapper".

The intelligence and grace of these huge animals never ceased to amaze me. An elephant helping to build one of our roads, or a retaining wall, would move up to a boulder with almost cat-like grace, with its ears flapping steadily back and forth, while the stone-masons fixed chains round it. Then, nudged by the feet of the mahout, would move deliberately into position. The trunk would reach out and curl round the chain, then tuck the fat, soft endrope into the great jaws. Then the elephant's huge feet would shuffle into a good hauling position. The trunk slid back down the chain, like a dark grey python, and wrapped itself round the links so as to hold it at just the right angle. Next the animal would lean well back to put its whole weight into a heave. Then just as it gave a mighty tug it would screw up its eyes, tight shut, just like a child. With one smooth jerk of its whole body the boulder was sent rolling down towards the building site. On another occasion the elephant's massive head would be raised so that the boulder would swing on its chains between the forelegs. Then it would move silently and steadily forwards, guided by its handlers, to place its load exactly where directed. The chain would be loosened, sometimes by the elephant itself, and carried in its mouth back to the quarry, ready for attaching to another load.

Elephants start work early, at dawn, and stop soon after noon when they go to collect their fodder and bathe. It is a fact that stall fed elephants will refuse to do any work after 12.30. Elephant Bill (J.H. Williams again) told of how the animals at Rangoon saw mills flatly refused to place the timber they were carrying on the stack, but just dropped it, when the whistle blew for the mid-day break. Bamboos are a favourite food and elephants need up to 600 lbs of fodder a day. Generally the trained (but not stall fed), animals get this by tearing branches and foliage from adjacent trees and shrubs. An elephant is capable of working for many years. If it remains fit and healthy it may work for father and son. It is much better than a tractor for certain jobs, being capable of negotiating steeper slopes and broken rocky terrain. It can enter the jungle and select its load with far less damage and disturbance. It is completely silent while working and discharges no pollution into the atmosphere. Also it does not need any expensive imported spare parts or fuel. The 1941-45 war in Burma proved that elephants can work in places and traverse land where bulldosers and mules cannot go. They often rescued heavy vehicles sunk in mud.

Elephants hired for work on Spring Valley were usually associated with local Bhuddist Temples. We hired them by the day, and with the elephant came its mahout and his assistant. There was usually some road building or other heavy building work going on somewhere on the estate. The old factory was burnt to the ground in 1943. The circumstances remained a mystery, but several other large factories in the district suffered a similar fate. Some said it was the work of Japanese agents. Whatever the cause it was decided to build a new factory high up on the estate close to May Mallay bungalow. It was considered that climatic conditions up there would enable manufacture of better quality teas. Meantime the little factory at Kottagodde was hurriedly developed and did sterling work, producing immense quantities of tea far beyond its theoretical capacity. This was thanks to Jack Cranfield's systematic tackling of the problem, helped by just a couple of planters who had stayed on long after retirement was due. I was away on war service in India and did not return until 1946. By that time, thanks to Lease Lend from America the new factory site had been prepared and the building was going up.

Several elephants were employed in connection with the work of construction and one of their last tasks was to move a 250 horse power Crossley engine, in its crate, up to the new factory 2,500 feet through a distance of 2.5 miles. This took nearly a month of patient work. No tractor available was capable of the task. When the load arrived at Demodera railway station it was transferred onto a specially constructed cart and hauled over the 8 miles from Demodera, taking about a week to negotiate all the sharp bends and to climb nearly 1,000 feet to the site of the old, destroyed factory. From this point the elephants took over. Two animals in harness hauled from the front and two pushed from the rear, using their great heads as buffers. In the narrower places the hauling animals had to pull in single file, one behind the other, since on several occasions the outside elephant was in danger of slipping down the "khud". Later teams of bullocks were used to steer from the front, while the four elephants took it in turns to push; two pushing and two resting. At each overnight halt large blocks were placed under the trolley wheels to prevent the load from running away. A guard was posted. The first six hairpin bends were successfully negotiated, but the seventh, at Pudumallay, presented a real problem. It was backed by solid rock and room for turning was almost non existent. In fact no vehicles turned here. They continued upwards past the "hairpin" for a hundred yards or so, to a small promontory where a turning circle was made. Vehicles went on up to this place, turned round, then returned down the short stretch to join the upper road at the "hair-pin". This presented a major problem to the engineers responsible for the transport of the engine. They were unwilling to proceed up to the turning platform because they feared the elephants would be unable to control the load on the short return section, as it was very steeply, downhill. They said there was a great danger that the load might take charge and plunge out of control to destruction. A halt was called, the trolley wheels were double chocked, and the elephants went off for a well earned week-end rest. By the next monday the engineers were still undecided, and arguing about the various possibilities. Eventually the elephant "muthalali" (headman) came forward and said "don't worry, the elephants will turn the thing round". Thereupon they went to work and, by inching the cart forward in minute movements one end, and backwards in tiny jerks at the other, and even lifting one end, then the other, momentarily off the ground, they very gradually turned it all through 325 degrees, while the engineers were saying it could not be done. The rest of us held our breath until it was sitting reasonably safely on the upper road. Thereafter very slow but sure upward movement resumed for the remaining mile until, at last,the engine was installed in triumph.

The harvesting of the tea crop is known as "plucking", and requires more labour and time than all the other works added together. In Ceylon (unlike Assam),the bushes ''flush'' throughout the year. Depending on the season, aspect, elevation and "jat", the pluckers must remove the "flush", (crop) from each and every bush every six to ten days. So, given an average of ten day rounds,this means that forty-five pluckings need to be completed every year. The only exception will be the quarter or so of the acreage out of production for three to five months following pruning.

"Good tea is made in the field", and "You can't make good tea from poor quality leaf" are familiar sayings amongst planters, which goes to show the importance that the quality of leaf has towards the character of a tea.In addition to a good standard of leaf it is very important that the leaf is all carefully handled. If it is crushed in the hands of the pluckers, or crushed in the baskets and open weave bags on its way to the factory, the tea will become heated and start to ferment prematurely, and so be useless for making into a quality tea. To overcome this problem pluckers have their baskets weighed several times daily. The weighed leaf is despatched to the factory as quickly as possible, by wire shoot, or leaf lorries into which it is loaded on racks so the weight of one sack does not fall on any other. On arrival at the factory the withering labourers are waiting, ready to handle the crop immediately.

When a field which has been pruned produces adequate good new shoots it is "tipped" by a special gang of "tippers" who use knives or shears to get an even "table" or "slope" throughout the field. Thereafter the same gang will repeatedly work through the field every few days to harvest the new growth. This young "tipping" leaf is usually kept separate in the factory, since the first few months crop is likely to give thin liquors of poor quality.

Most Divisions and Estates organise their pluckers into three gangs. The tippers and Young Field pluckers form the first gang and work in fields of up to a year or eighteen months from pruning. The pluckers in the older fields form another, and a third gang, often consisting of older women and men, will be employed at plucking fields due for pruning during the next six months or so. The Young Field gang is made up from the younger women who have very nimble fingers, and are quick and intelligent. It is a pleasure to see these women at work in their brightly coloured clothing, laughing and chattering as their supple fingers take the flush with remarkable speed, and throw handfulls over their shoulders into the baskets slung on their backs. They stick to their work through all kinds of weather, and it is usually the manager or superintendent who has to order them home when the conditions are really bad.

I have spent many hours with the pluckers in every one of the seven divisions on Spring Valley. To my mind those Young Field pluckers on May Mallay during those pre war years were easily the best of all the gangs. I can picture, as if it were yesterday, pluckers taking up their rows early on a cold, bright morning, at the Cannavarella Gap, the ridge where my tea bordered onto Cannavarella estate at an elevation well in excess of 5,000 ft. Standing on this ridge, at that hour, there was a view to the East, across the Low Country jungles,.to the sea which shone like a bright silver ribbon as the sun rose out of it. It was all of sixty miles away. The road to this gap had just been completed and joined the much rougher affair leading down through a series of hairpins into the district of Namunukula. The tracing,cutting and building of the road on the Spring Valley side, with its many bridges, culverts and retaining walls, was one of my first tasks on Spring Valley. Over the years I would see the tea yield grow from 450 pounds per acre to 1,500 lbs, but that was yet to come.(Note The yield of tea is expressed in pounds per acre of the finished product which we put into our tea pots. This is approximately four times less than the weight of green flush harvested by the pluckers.)

Turning towards the West the sheer beauty of the scene was enough to affect anyone deeply. Immediately below was the deep saucer shaped valley dominated to the North by the towering jungle clad peak of Namunukula mountain. The steaming plains round the small town of Badulla, the provincial capital, lay nearly 4,000 feet below the edge of the saucer on whose rim I was standing. All round the lip of the valley was spread the brilliant green of the tea, like a gigantic green girdle. This was the season when rain and sunshine alternated, when the miles of bushes, usually a dark green, rather like laurel, were covered with tender lettuce green new shoots which would soon fill millions of teapots waiting all around the world.

It was at this season that we young planters best loved to plunge down into the tea to meet the gay laughing chattering army of men, women and children, often many hundreds strong,and extended in a long line abreast. This army of pluckers, baskets on their backs, suspended from cords across their brows,worked with flying fingers through the flushing tea. Everyone of them, from twelve year old learners at their mothers sides, up to and including their grandparents, worked with an intensity of purpose. The crop must be harvested while in the best condition; while the young shoots were still tender. This way the next crop would develop sooner and be of better quality. Everybody on the estate knew the effects of harvesting the leaf too late. The crop would be less; the quality would not be so good, and the price paid at the London market would compare unfavourably with adjacent estates.

Everyone took a great interest in maintaining the quantity and quality of the crop. After plunging into the tea the Manager, Peria Durai or Sinna Durai, would work on foot past the long line of pluckers, every one of whom we knew by name. As we passed them we exchanged here a jest, there a smile, and with another a quick enquiry regarding some sick or aged member of the family. For years we shared the heat of the long day with them, listened to their complaints, helped to settle their quarrels impartially, laughed at their marryings, grieved at their buryings; giving them, in fact, everything to which their feudal loyalty to us entitled them. We were privileged persons and we knew it. We also knew that with those privileges we assumed the full obligations which went with them. We all lived close to nature and understood what nature could do; aware of what is so often forgotten or overlooked in a western society propped up by taxes, subsidies, grants and other artificial means; that to harvest the crop well at the right time, was likely to bring prosperity, and to lay the foundations for more prosperity in the future. Whereas loss of crop would have the opposite effect, leading even to misery, loss of employment and, eventually, starvation. This was something our supposedly ignorant and primitive people recognised so much better than so called "civilised" Western Society where Greed is more often the incentive.

It is not my intention to suggest everything was perfect; far from it! There was much to do to improve conditions. My fellow planters would, I feel sure, have been reluctant to think of themselves in the terms I have mentioned above. Nevertheless nearly all of us had the welfare of our workers at heart. We fought the Colombo Agents and the proprietors to secure better conditions for them. What I have written in the last paragraph may be greeted with cynicism if it is read by any of those brought up in the post war years when it has been the fashion to denigrate the achievements of the past.

The Sinhalese peasant had no wish, and no need, to labour on estates, so Major Skinner's prophesy made in 1840, that it would indeed be:-

"conferring a blessing .... to be the means of removing some of the panting, half famished creatures, from the burning sandy plains of South India, to such comparative paradise; benefitting not only them, the colony, the individual by means of whose capital they would be brought here, but also our own native Singhalese people inhabiting the margin of this wilderness".

Even at the present time (1991) after a threefold increase in the village population accompanied by much unemployment, work in the plantations appeals to very very few of the Sinhalese labouring classes. The estates continue to be worked predominately by Tamils. But as prophsied, "the Sinhalese inhabiting the margins" of estates benefit as suppliers of goods and services to the industry and its workers.

Changes in the relationship between plantation management and labour were on the way, and began to manifest themselves in the early days of the war when the number of PDs and SDs suddenly fell sharply, together with the amount of supervision. Agitators were able to enter estates and started to foment trouble amongst the workers.

Bridge building, 1st Division, 1938

CHAPTER SEVEN

THE END OF AN ERA

"When tempers get heated and things are fogged, adjourn for tea.
It is a priceless prescription."

Lord Curzon, when British Foreign Secretary (1922)

Before the onset of the monsoons the mornings would often be crystal clear and cool, giving a freshness which made it a pleasure to get up and go out with the dawn. Later in the day however, as the unstable mass of air in and over the mountains heated up, clouds began to appear. These clouds soon began to thicken, and to tower upward. Their threatening black undersides would approach nearer and nearer. Black shafts of tropical cloudbursts could be seen against the distant slopes, while peals of thunder grew heavier and more frequent. Flashes of lightning played in the distance, sometimes flickering for many seconds so that the mountain crests showed sharply in silhouette.

These storms were unlike any I have experienced elsewhere. As they came closer the thunder reverberated and the lightning blazed. The tropical rain could be seen approaching like a grey hissing wall, blanking out and making everything it had passed invisible. Suddenly the rain would be upon us, hammering down with great force and bouncing up off the ground before forming rushing little rivulets. As the very core of the storm swept upon us there would be unbelievable displays as the constant flash of lightning and crash of thunder reverberated back and forth between the mountains. Jagged lines flashed down to earth and burst over our heads like an artillery barrage. At times the discharges came so close that they produced an almighty crack like a giant whip so that we ducked as if under rocket fire.

In April 1939 we had a series of such storms of almost terrifying intensity when, for three or four afternoons in succession, there were downpours which darkened the day almost into night. On each occasion these produced between three and five inches of rain in less than that number of hours. The damage to roads, culverts and bridges was considerable and the mountain sides erupted into sheets of falling water. When the rain ceased, as abruptly as it had begun, the roar of the waterfalls was loud all around. I was out in this atrocious weather supervising, the run-off of the water and seeing to the clearance of landslides and blockages. The result was a bad cold followed by a high fever. The doctor came; took one look at me, and transferred me immediately to his bungalow at Bandarawela. I don't remember much of that time except that Dr. Vesey Adams and his wife were extraordinarily kind and painstaking and looked after me as if I was a member of their family.

I occupied their spare room for about a month having developed a bad go of pleurisy whilst in a generally rundown condition. Vesey Adams would look in to see me several times during the day as well as during the night and, as my condition improved Mrs Adams provided me with a specially delicious light diet which included excellent wines with every meal. It was only much later that I came to realise how much I owed to the Adams. Looking back I often wonder if I showed proper appreciation cf their kindness.

Those thunderstorms could cause tremendous damage. On one occasion years later, on Kottagodde, when I was ill again (but this time had the benefit of my wife, a nurse, and a nurse from the Ceylon Nursing Association to look after me in the bungalow), a freak thunderstorm was accompanied by a whirlwind and a hailstorm. Over one hundred fine grevillea and albizzia trees were completely uprooted close to the bungalow, and hailstones the size of pebbles drove right through some of the corrugated iron roofing sheets on the stores and lines, perforating them like a collander. It was not uncommon to see patches of 30 or more tea bushes killed by lightning and to have fuse boxes blown to bits. There was a lot of ironstone in the rocks covering the hillsides and this attracted the lightning. During a severe storm static electricity could be heard by its continuous humming.

One fine clear morning during the following autumn, probably in September before the onset of the North East Monsoon, I went out to find the whole world seemed to be a moving mass of white, like a single continuous rushing cloud. On closer observation it was seen that the movement was close to the ground - up to a height of about eight feet - and was going up and down hill, across rock and stream, following the contours. It was a mass migration of butterflies. The swift galloping flight of these migrators, in an immense swarm of what must have been many millions, was in striking contrast to the weak hesitating flight of many species Of lepidoptera. The flight was from the East towards the West and the kanakapillais and kanganies on the estate said they were all going on pilgrimage to Adams Peak, the holy mountain.

L.G.O. Woodhouse in his book "The Butterfly Fauna of Ceylon", mentions "the fascinating problems of migrations of butterflies" and says that many species make long flights similar in nature to those of migrating birds, "making flights covering hundreds or at times, thousands of miles........ thousands or even millions flying simultaneously. In some places these flights are sufficiently regular to have become the bases of legends and superstition among native races". Ceylon is reported to have over 60 species regarded as "flighters". Although I saw flights fairly regularly in later years, never afterwards did I see quite so great a migration. It went on for several days.

When I had been nearly a year on Spring Valley Jock Sutherland asked me to take on the job of Honorary Secretary and Treasurer of the Anglican parish covering the northern half of Uva Province. I probably felt it would be impolitic to turn down Jock's request since he was the senior church-warden as well as my boss on the estate. Although I agreed rather reluctantly to the proposal I was to find the extra work interesting and it gave me rewarding experience of Sinhalese and Tamil life outside the planting sector. The parish was scattered over a very wide and rugged area extending to the seven planting districts of Badulla, Demodera, Passara, Madulsima, Lunugala, Namunukula and Moneragala. There were four churches; St.Mark's, Badulla; St.Peter's Lunugala; St.Barnabas, Passara, and Christ Church, Demodera. All except St. Marks had been built with funds from the estates and planters donations. About once a month, or whenever he could spare the time, the vicar held services at the planting clubs in Madulsima and Namunukula, whilst at Moneragala a room at the Rest House had to do duty for the occasional service.

Twin Velocette 7.90 h.p. 1919 model.

Woleseley Hornet Daytona Special, 1934 model

Elephants transporting a heavy load.

The church of St. Mark's in the centre of Badulla town had an unusual beginning. It was built from funds donated by local people of Bhuddist, Moslem, Hindu and Christian faiths, in honour of a Major Rogers who had killed over 1,300 elephants in the area. Today Major Roger would be vilified as a callous brute for destroying such wild life. In his life-time he was regarded with awe and wonder by the villagers he saved from the destructive depredations of the elephants. Writing in the 'Tropical Agriculturist' in 1890 the late A.M. Ferguson C.B.E., stated

"It was Rambukpota Dissawe who brought up the idea of a Christian place of worship in memory of the late Major Rogers. He and his fellow chiefs, all 'Bhuddists...... contributed each a month's salary towards the object and the minor Headmen also according to their means. The Dissawe, besides his subscription, gave nearly all the timber for the building. Packir Thamby, constable, as head of the Mohammadens, got large subscriptions from that community, and the Chetties also followed suit liberally The foundation stone was laid in 1846, and in a bottle was put a coin of the date of 1750, by the late Mr Solomans. The various officials who were connected with the building of the church were Messrs. Mercer, Braybrooke and Bailey. It was distinctly understood that the church would be open to all protestant denominations, but how it went over to the Episcopal Church this deponent sayeth not, and it was convenient that no questions should be put."

It looks as though there was some sort of row about who should use,the church but in my time, so far as I am aware,it was certainly used for weddings and funerals by members of various denominations. The North wall of the nave bears a plaque dated 1845, with the following inscription;-

> "This Church was erected to the Honour of God
> In the Memory of
> Thomas William Rogers,
> Major of the Ceylon Rifle Regiment,
> Assistant Government Agent & District Judge
> of Badulla
> By
> All Classes of his people, Friends and Admirers.
> He was killed by lightning at Haputale,
> June 7th 1843
> In the Midst of Life we are in Death",

Major Rogers is still spoken of, in Uva, with considerable respect. It was he who opened up the very first estate in Uva (in coffee), at Ridipane. That estate is still called "Major Totem" by the labour force.

In view of the greatly changed attitude towards the killing of wild life compared to earlier times, another short note about Major Rogers seems appropriate. He was probably a good example of the young military officers, who in the early days of the British occupation, were not called upon to fight, but gave a life of almost complete devotion to the Civil Service of Ceylon. He landed in Trincomalee in 1824 as a young man in his twenties, and joined the Ceylon Rifle Regiment as a second lieutenant. In 1828 he was appointed Commandant of Alupota, a small fort established during the Uva Rebellion to guard the foothills to the east of Passara. He spent six years in this wild roadless jungle station. The whole district was overrun by herds of wild elephants which destroyed villagers crops and were a constant terror to any travellers. They took a fearful toll of human lives and some said the elephant population exceeded the humans. They were like the wolves of Europe in earlier times, but much more dangerous. As civilisation began to conquer the wilds, they had to be overcome, and a price was set on their head. Rogers soon established a reputation as a mighty hunter and many legends grew up about his name.

In all his long warfare with wild elephants Major Rogers only came to grief once; that was on 29th December 1841. He had already shot several elephants that day and was following another through the forest. The animal suddenly turned round and, catching him in its trunk, flourished him over his head as if he were waving a flag. Rogers hung helpless awaiting his fate. The elephant carried him to a small stream and dropped him on the sloping bank, trying again and again to crush him with its head, all the time trumpeting its fury. Luckily Rogers found the ground in his favour and each time he saw the great head start to come down towards him he managed to roll a few feet down the slope, avoiding the worst of the blows. When they reached the bed of the stream the animal tore at his clothes. Fortunately these were rotten with sweat and tore from his body leaving him almost naked. Major Rogers pretended to be dead. Suddenly something startled the elephant. It seemed to become suspicious. Rogers lay quite still and the great beast stepped right over his body as if with deliberate care, and began to move away trumpeting furiously as it went. Rogers was picked up by some of his villager friends who had watched the whole incident from the safety of branches in tall trees. His left shoulder was dislocated, his left arm broken in two places, and he had two bad wounds on his right side. He was fifty miles from home in a dark and almost trackless forest. His followers picked him up and carried him all the way to Badulla where they arrived a day later. He made a good recovery and was none the worse for his adventure. He made a deep impression on all the country people who believed he bore a charmed life. " Nothing in the world can harm that man", they said. "If he dies it will be by lightning". The modern map of Uva is to a great extent Roger's work. He won the respect and love of all the people. He went among them constantly until he understood all their problems and their hopes.

Rogers took refuge from a sudden thunderstorm,on 7th June 1843, at Haputale Rest House. After a time the violent storm appeared to be over and Rogers stepped outside. Suddenly there was a blinding flash of lightning followed by a shattering thunderclap. The pandal over the rest-house entrance was split down the middle, The coolies and horses in the back verandah and the outhouses were all struck down, but not seriously injured. Rogers fell dead. He was buried in the little old military cemetery at Nuwara Eliya. This cemetery still contains the tomb. It was split in two by lightning even after he was dead. The cemetery is situated on a small mound just behind the golf clubhouse. Rogers was a worthy representative of the old Civil Servants to whom Ceylon owes so much, but who are now very seldom remembered. With this fine example of service how could I not take an interest in the church of St. Marks in Badulla? The churchyard has, or had, many beautiful trees, and contains some interesting tombs, amongst which are a number of planters and their wives who I knew very well. There are many ways a planter could come to grief and this is well illustrated by the tombstone erected in 1880 which reads:-

"William Bennison Jnr.,
of Mausa Estate Hewa Elia
aged 23 years
This monument is erected by his brother
planters of Madulsima and Hewa Elia
as a token of their respect for his memory
He was shot by his Appu* during dinner.".
*cook

Not long before I took on the church job the very popular, immensly tall, grey-bearded Father Waltham had retired after over forty years as vicar. Jock had recruited a new padre whilst he was Home on leave. This man proved to be a dismal failure, being more suited to suburban vicarage tea parties than to a vast parish in such a wild and outlandish district. When I took over as Hon Secretary negotiations were under way to send this man Home. Parish meetings were held at the vicarage and tended to be lengthy. Amongst the church council members I remember Sentence Smith of Telbedde, C.Aubrey Clarke of El Teb, Passara, Major Playfair from Madulsima, and Messrs Lisk and Pinto of Badulla. The church owned glebe lands in very remote places, and some buildings in Badulla. There was always great difficulty in collecting the rents, or whatever these dues were called. Most of the income to maintain the churches and pay the incumbent came from an acreage cess, voluntarily agreed by the estates. Estate Superintendents and Assistants were also canvassed annually for subscriptions. Some gave very generously and several of the most handsome donations came from members of the Scots Kirk who never attended any of the services.

Our European padre eventually left for England and while his post was vacant I was asked to take charge of the vicarage furniture. This was a great boon as the Wilsons were back from furlough and had taken all the stuff they had left with me to their new billet at Ellawatte. In their place I now had comfortable chairs, sofas, carpets and other articles from the vicarage. When Father David the new vicar, was appointed he lived in his own house so I continued to look after the church property. Whilst the vicarage was empty I had it surveyed. It was found to be quite unfit for human habitation owing to termite attacks and dry rot etc.

After the war the house was patched up and put into use again as a vicarage though I suspect it was still not really fit for human habitation. Planters were not, in general, good church goers but everyone, just everyone except the wives would turn up to see a colleague sent to his rest. Due to the climate this had to be done within 24 hours of death.

Jock Sutherland had a reputation (undeserved in my judgement), of being a bit of a skinflint. He had short arms and long pockets they said. If buying all and sundry rounds of drinks in the club is a good measure for generosity then they were right. I found that when it came down to brass tacks Jock could be extremely generous. I suppose that, like most good Scots, he spent his money wisely. His name, invariably headed the list of generous subscribers to the church, the Salvation Army, The Earl Haig Poppy Appeal and so on. He also was known to have helped some planters when they were in debt. The district sported several planters of independent, forthright and determined character, such as were seldom met in later years. Charles Ruxton of Hopton Estate, Lunugala, was one such. He ran. a very smart estate and kept his Assistants well up to the mark. Walter Ogilvy used to tell a story of how he once got the better of Ruxton. Walter's bungalow adjoined Ruxton's, or rather the gardens did. Walter used to rise very early in the morning and always had a large bowl of porridge for breakfast. This he sometimes consumed in the garden, to enjoy the early morning air as well as to see that the labourers were heading for their work places in good time. On one occasion at 5.30 a.m. he noted that the milk, supplied by Charles Ruxton's cook appu, was well watered. He went into the bungalow and reported the fact by telephone to the appu. Charles Ruxton overheard part of the call and asked the appu's for the details. He tried to get his own back a few days later by accusing Walter's appu of stealing his tomatoes, and demanded he be punished. Walter requested evidence of his appu's guilt and Ruxton said he had seen his footprints. How did C.R. recognise the footprints asked Walter? This stumped C.R. and no more was heard of that matter. However Charles Ruxton began to visit Walter's bungalow in the very early hours, 3 or 4 am, to chase him out to work. Walter decided he would cure this annoying habit and bought a parrot which he taught to say "Peria Durai varutha" (The Manager is coming), whenever anyone approached.

On another occasion Walter (WTO), and his fellow SD, Norman Dru Drury, (NDD), had arranged a very early departure, on a "Perunal"* when there was to be no estate work, to sit up over a calf killed by a leopard. The leopard had been terrorising the labourers at the bottom end of the estate, coming in from the jungle at night and regularly taking dogs and poultry etc. Charles Ruxton telephoned at the last moment asking NDD to take his son into Badulla to watch a rugger match. NDD refused saying he had laid on full arrangements to go shooting with WTO. Ruxton was furious and ordered both the men to report to him in his office immediately. They both went the five miles from NDDs' bungalow on WTO's motor bike. They were told "You fellows have too much spare time and wont take my son to rugger so take these two extra check rolls to keep you busy working in the evenings"!

Two other characters, my seniors by about 15 years, were J.A.F. Wallace and A.K.Will (brother of Percy Will). There was a story about these two which took place when they were still SDs and bachelors. The great Andrew Young, the chairman of many Ceylon plantation companies based in Mincing Lane, was on a visit to Ceylon and was entertained to dinner at Ouvah House, the then headquarters of the Ouvah Ceylon Estates Limited. in Badulla. The host and hostess were the company's Ceylon manager, Joe Scott and his wife. The superintendents and assistants of all the company's properties were present with their wives,(though not many were married) Jack, Wallace and Alec Will were near the bottom of the table being mere juniors. During the meal Joe Scott was heard to suggest that they should all repair, after coffee and cigars, to the big-top tent on the "maidan" (an open space or plain) to attend a rare performance by an itinerant circus.

Later in the evening the Party went off in a body to the circus where they were directed to ring-side seats. Alec and Jack made excuses saying (truthfully) that they had already seen the acts and intended to return home. The party were enjoying the performance which, though rather inexpertly staged by western standards, was a welcome and unusual event for most of those present. After a while the major domo announced there would be a spectacular act by camals and their riders, whereupon two rather mangy beasts entered the ring ridden by two men dressed as Arabs. They circled the ring at a trot with their mounts clinging on and occasionally rising, seemingly with difficulty, to a more or less standing position. Every time the camels Approached the side of the ring occupied by the evening-dressed Europeans the riders seemed to be urging their beasts to swing round and go astern, so that their hind quarters overhung the ring causing the spectators to press themselves as far back as possible to avoid contact or contamination. Suddenly Andrew Young turned to Joe Scott and announced loudly in his broad Scots accent "Do you Ken, Scott ? One of yon laddies is mighty like wee Will". Immediately the camels made rapidly for the exit. Realising they had been rumbled, but with the intention of establishing an alibi, Alec and Jack dashed to Ouvah House, leapt on their horses and rode hell for leather to their respective bungalows on. Glen Alpin estate. Presumably the animals were used to the route and knew the way well in the,dark. Alec had farthest to go - nearly six miles with a rise of over 2,000 feet - and in his haste arrived at a gallop on the bank above Grahamsland bungalow. His horse put on all four brakes to stop going over the edge and Alec shot over its head to land with a crash on the corrugated iron, lean-to, roof of the kitchen. This at least had the benefit of waking his servants from their slumbers which gained him speedy access, just in time to answer the telephone. Joe Scott was ringing to enquire if he was at home.

The Glen Alpin Manager's bungalow where Joe Scott then lived,was about 1,000 feet above the road-end where he garaged his car. When he and his wife visited Ouvah House, or the Uva Club in Badulla, they would return by car as far as the garage where horses and carrying chairs would be waiting. Joe and his fellow male guests would mount the horses and canter off up the zig-zag track,
*holy day.

84

leaving the ladies to follow in the carrying chairs, their way lit by hurricane lanterns. This was a good opportunity for the men to down a few whiskies before the ladies arrived.

By the time I came to the district there was a motorable road right up to Glen Alpin bungalow. Percy Will had taken over from Scott who I first met when he was my company chairman in London. I found him charming. Later on our children used to refer to him as 'Uncle Scottie'. In due course Jack Wallace became Manager of the Ouvah Company, and later still was one of my directors. He distinguished himself in the second World War, was taken prisoner at the time of Dunkir,k, was involved in several escapes and spent a long spell in Colditz. He and his charming wife Marian used to look after our daughter at half term when she was at the Royal Naval School, Haselmere. Alec Will died in harness whilst superintendent of Elladallua estate, Badulla. There is no doubt that the planters world was male dominated. In those days the number of bachelors far outstripped the married men.

I used to visit Badulla about once a week. Usually to play a very poor game of tennis, or hockey and cricket according to the season. The Sinhalese were very nippy at hockey and we had some good games. Tup Grant-Cook's side beat the Colombo Cricket Club that year when they came up at Easter, but I think it was mainly due to Tup's wily tactics and local knowledge. We managed to get their best batsmen out after a very liquid lunch break. Our opponents seemed barely able to stagger to the crease. Uva then fielded about the best rugby XV in the island. The home matches always drew a big crowd of planters and wives, and the locals turned out to cheer for our side. The rugger matches were always followed by a good party and we often took our guests from the opposing team back to our estate bungalows in the early hours. Somehow we managed to negotiate those tortuous hill roads without disaster. Several of our men - usually sons of planters - seemed to have a natural aptitude for games, as if born with a rugger ball, or hockey stick, cricket bat, or shot gun in their hands. Sandy Richardson and Donald Marley were "naturals" at any ball game, and were a delight to watch. They could have gone far if they had taken their games seriously. Donald Marley had a twin brother. No-one could tell the difference between them and when they were at school at Marlborough they were said to have changed places whenever it suited them. When babies their parents had to tie a blue ribbon to one and a pink to the other and they became known as'Blue and' Pink Marley. Walter Ogilvy played a very stylish game of tennis and was a beautiful golfer. Gerald Du Pre Moore (Dopey) was in a different category. He was extremely short sighted, and yet had perfect timing. He didn't have the flowing grace of the others but had tremendous gifts and determination. I have been out shooting teal with him and he has asked me "What are those birds?" when a flight of duck went over just out of range. When they came back a bit lower down he got left and right, bringing the birds down stone dead.

"Les Oiseaux" were generally notable by their absence. The "fishing fleet" from Australia failed to materialise, or had all been caught before they got Up-Country. However with the increasing threat of war a few planter parents brought their daughters out to the island "for safety". The two Pickering girls, Phoebe Grant-Cook, and the Australian niece of Bill Adams, Kathy Adams, are the ones I remember best. Bob Carswell, a very level headed and slightly older young planter took unofficial charge of these girls and made sure the wilder ones amongst us behaved properly towards them.

After tennis, hockey or cricket - I was not a rugger player - we would shower and change into slacks, long sleeved white shirts and ties, then repair to the King's Theatre, the recently opened bare tin shack of a cinema. Long sleeves and covered legs were essential for protection against marauding malarial mosquitos which became active after dark. Mosquito boots and anti mosquito cream were also advisable and the ladies tended to wear trousers or long dresses. When interest in a film flagged

Spring Valley old factory: elev 3,000ft..
Destroyed by fire in 1943, possibly as a result of Japanese action.

Spring Valley new factory, completed in 1949. Elev 4,900ft.

characters like Noel Dyson-Rooke and Denis Johnson would keep us, and the local cinema going clientele, entertained by their outrageous mimicry and comments in the vernacular. This would sometimes develop into a hilarious two way flow between us in the rear seats and some of our workers in the sixpennies. When my transport failed, not infrequently, Dopey Moore would take me riding pillion. He used to put the fear of the Almighty, or the Devil, into me by speeding down the hills in hairbrained style, leaving culvert marking stones scattered behind as he scraped his foot-rest against them while cornering at speed. I was not averse to driving fairly recklessly myself, but did not enjoy putting my life in the hands of someone else. Traffic on those roads could be counted on the fingers of one hand. In fact I could usually tell who was approaching by the tone of an engine, and, at night, approaching headlights gave ample warning of any oncoming vehicle. It was the nature of the roads, their twisting and turning, and the hairpins and the"khuds" which were the hazards.

The Uva Amateur Dramatic Society put on the occasional play at the club. I think that in 1939 it was Emelyn Williams "Night Must Fall". It was a huge success. Bill Nichol of Cullen Estate, made a fine judge. Noel Dyson-Rooke was the murderer, Budge Birkett then on Sarnia Estate was a detective and, I think, Phoebe Grant Cook was the victim. I suspect that when I was able to coax my Wolsley Hornet Daytona Special sports car to cover the journey home from Badulla to May Mallay late at night, I was very unpopular. The roar of the exhaust would have disturbed the Cranfields on Kottagodde as I twisted and turned up the valley and went below their bungalow. Then I passed below the Sutherlands. I usually had to stop to add water to the radiator, and to cool the boiling engine before attempting the last, exceptionally steep, three miles in second gear. The echoing roar of the engine, from side to side of the valley, would not have disturbed the labourers who shut themselves tightly into their line rooms and could only be awakened by an earthquake. I learned later, from personal experience, that the noise could be clearly heard at Big Bungalow. Such is the inconsideration of Youth. Fortunately for my pocket these expeditions were rare. It was far cheaper to go off on a shooting trip to the Low Country jungles when time or leave permitted. It was a great experience too. The opportunity, when it occurred,was usually at the time of Thai Pongal in January, or Adi Poossy in August. At the time of the former festival we would be after snipe and teal. In August it would more likely be pigeon of which there were many varieties in the Ceylon jungles.

The best snipe and duck shooting involved a four to five hour journey towards Batticaloa or Pottuvil on the East coast. Whenever possible we would make up a party of about four men and two vehicles. A second vehicle acted as a sort of insurance in case of breakdown. A very early start at about 3 or 4 in the morning made it sure that we could get down to the flat low country jungle, near Bibile or Moneragala, by dawn when there was a chance of seeing elephant, leopard, deer or even a bear on the road. The animals often came out of the jungle at night, when it was wet, onto the roads to avoid the dripping forest. The first vehicle down the road in the morning would scare them back into the jungle.

On the way to the coast we were likely to make a diversion down one of the Irrigation Department's earth roads to a large lake, known as a "tank" to bag a few whistling teal, cotton teal, or even some gargeney or pintail. The first two were resident varieties whilst the last couple were migrants. These ancient tanks, mostly rediscovered and restored by the British, were extraordinarily beautiful and supported (and still do) immense populations of birds of too many varieties to mention here. In a letter Home about that time I tried to describe Rugam Tank saying "You can see more birds here in half an hour than you you will see in a whole year at Hasketon" (our home). A few of the villagers would appear from nowhere for a chat, and some of the small boys would volunteer to act as retrievers. We always took some small packets of tea with us and these were gratefully accepted as payment, apparently being more appreciated than cash which had little value due to the absence of markets. Most of the villagers in that area showed distinct signs of suffering from malaria, being pale and lethargic. Many of the small children,who ran about naked, had enlarged spleens caused by that disease. They were always courtly and polite. The Sinhalese peasant was one of natures gentlefolk. Only those from the cities and in contact with "civilisation", tended to loose those qualities.

Continuing on our way to the coast we would check our rooms at the primitive Rest House on the sea shore then, in the cool of the late afternoon have a go at walking up snipe along a lagoon edge, or in paddy fields, returning to the rest house soon after the sudden onset of a tropical night. After a dip in the sea we would sit out under the stars,yarning and enjoying our "sundowners". The night sky in the tropics has to be seen to be believed. It has more colour than the skies of the temperate zone. We could see the Southern Cross low down to the south, and to the north, also low down near the horison, was the Pole Star. Either the Plough or Casseopea would be visible but only very rarely could both constellations be seen. A meal of fish, crabs or prawns - fresh out of the sea - perhaps followed by curried snipe or duck from our bag, would satisfy the inner man. Then early to bed under mosquito nets. The cool sea breeze had kept these pests away whilst we sat on the beach, but they lurked everywhere just a little way inland. Lying in bed we would hear the plaintive calls of curlew and plover, the Tchk-tchk-tchk of ghekkoes in the thatched roof and the muffled sound of breakers on the distant coral reef, before being lulled into deep and satisfying sleep.

A break of two or three days , a complete change from life on the tea estate would be a marvellous tonic. The only traffic was the occasional bullock or buffaloe cart, sometimes accompanied by the drivers high-pitched calls urging his beasts to keep moving at their steady two mile per hour. There was no mail, no telephones and no newspapers. We took most of our stores with us, including, if we were wise, our own drinking water, beer, liquor, towels and bedding.

On the estate itself, there were jungle and spur fowl; which came into the tea from surrounding jungle late in the evenings and at dawn. There were the occasional mouse deer, and a leopard was known to inhabit the Namunkula jungles at that time. Below the estate were steep, terraced paddy fields. On the odd week-end in January or February, Dopey and I would take our guns and work our way all the way down to Badulla, shooting as we went. This was hard work. We frequently sank up to our knees in the mud, and the birds often fell far below us, or on the opposite side of the steeply terraced and narrow valley. By the time we reached the club in Badulla, where we had sent our horses, we needed plenty of cold beer to replace weight lost through perspiration. Then a quick ride home to one of our bungalows, a good curry lunch followed by an afternoon snooze. I never shot any of the jungle or spur fowl, or for that matter the mouse deer. These were too precious and the call of the jungle fowl was too attractive to destroy.

A few words about the festivals,which made it possible for us to take time off for those never-to-be-forgotten breaks, may be of interest. It was the universal custom for the workers to have one chief annual holiday lasting several days. This holiday was always an occasion for a big 'Swami Coombadu' (lit=God Worship). In Uva it was taken in the dry season when work was at its slackest, in July, as the name Adi Poossy (July Prayers) implies. The actual date was the subject of mutual agreement between the management and the workers as represented by their kanakapillais or kanganies. I think this demonstrates the depth of understanding between the parties. It was then that there was time for some of us to get away to the jungle, to the east coast, or to the bright spots in Colombo for those few who could afford or desired that sort of recreation. One member of the planting staff always remained on the estate to keep an eye on things, and to attend to any of the rows which tended to crop up between various factions, due to drink, women trouble and perhaps caste etc. Adi Poossy usually coincided with the Vael Festival for the god Subramanian, the second son of Shiva. There is a famous shrine to Subramanian at Kataragama, in the jungle near the South East coast. During July many of the families make pilgrimages there, particularly childless women who wish to change their fortunes in this respect.

Thai Pongal (lit= January boiling) appears to be a survival of the ancient sun-worship. It occurs on the date of the winter solstice, i.e., mid-winter in the Northern Hemisphere, when the sun is vertical over the tropic of Capricorn. The purpose of the festival is to welcome the return of the sun to the North of the Equator (i.e.India) at the time when the sun is needed to ripen the crops. It lasts three days. On the first it is the practice for everyone to visit and entertain each other all day. On the second day visiting is resumed and the first words spoken at a meeting are "Has the rice boiled?". To this the answer should be "pongarathe". (it boils) "Pongal" literally means "a boiling". Certain ceremonies with boiling rice are carried out by married women on this day. The third and final day is the "Madu Pongal" (Madu = Cow or cattle). On this day everyone makes much fuss of and specially feed their cattle, with much ceremony. In the evening dancing, singing and acting take place. The performances are notable for their obscenity as anyone who has witnessed them will know.

I was sometimes invited to attend shows put on by itinerant players visiting May Mallay. I will never forget one such performance. First a magician swallowed eight or nine solid and brightly coloured balls the size of ping-pong balls. We were invited to feel his bare tummy where the balls formed definite bumps below his skin. He then asked us to choose any colour we wished to have and regurgitated our chosen colour. He never failed to bring up the correct colour! The scene of these performances was very attractive. Several hurricane oil lights, and a few petromax type lamps would be hung around the open muster ground. There would be a clear space with a white sheet suspended at one end, to form a back-drop. A couple of hundred, or so, tamils of all ages and both sexes would squat round the edge of this arena. I would be given the only chair and a clerk, or senior kangany would be next to me to make sure I understood everything that went on. Various players would appear from the dark background to juggle, play, or whatever. The silhouette of towering hills behind were topped by a sky full of bright stars. On this occasion the most remarkable performance was one of "levitation". I will do my best to describe it. The actor, subject, patient, or call him what you will, was prepared in white clothing inside a small tent, with much ceremony and incantation. After a while the tent was slowly removed amidst loud drumming. The throb of the drums grew faster and faster, then gradually subsided until they gave a quiet, peaceful rythm. At this point the levitated subject was seen to be stretched out horizontal to the ground, stiff as a board. He had one arm outstretched towards the tip of a walking stick. The tips of his fingers just touched the head of the stick. There was no other visible support. I asked if I might approach for a close look. This was allowed but I was told on no account to touch anything. I satisfied myself that there were no strings or wires or any support other than the fingers on the walking stick. I have no idea how this levitation was achieved.. There were several other acts such as snake charming, and the well known mango trick. Here the magician produces a ripe mango fruit, covers it with a cloth, in full view of the audience then, after various songs and incantations removes the cloth and the fruit has thrown out a bud on top and roots beneath. The performance is resumed until, a few minutes later a young mango tree about two feet high is standing in the centre of the arena. These shows were great fun for everybody but as the years went by seemed to become rarer and fewer.

During that year it seemed that every week-end there was a ghastly crisis which caused shivers to run down one's spine and one's stomach to churn uncomfortably. These were brought about by Adolf Hitler, Benito Mussolini or Emporer Hirohito, or by a combination of them. After the Munich crisis there was a very great sense of relief at Chamberlain's "bit of paper". But this did not last very long. Hitlers' annexations were always "the "last". He always said he had "no more territorial ambitions", but within a few weeks he would be making more outrageous and unjustified demands. Appeasement had been the order of the day and the dictators had come to expect that they could get away with absolutely anything. We felt very much out of touch with events,in Ceylon, and exposed to the East where the Japanese were overrunning more and more of China, and were extending their propaganda and spies througout Asia.

Ever since Hitler had become Chancellor of Germany in 1933, and both Germany and Japan had left The League of Nations that year, tension had gradually built up. 1934 had disclosed the first photographs of concentration camps, and Hitler and Mussolini had held their first meeting. In 1935 Germany introduced conscription, condemned by the League of Nations, and entered the Saarland. Italy invaded Abyssinia, using brutal tactics and poison gas against the defenceless people. In 1936 Germany marched into the Rhineland to occupy it unopposed. In 1937 Italy made piratical attacks on neutral shipping in the Mediterranean, and left the League of Nations. Japan invaded deep into China, crossing the Great Wall, capturing Shanghai, and bombing Canton and Nanking indiscriminately, sinking an American warship in the process. In 1938 Hitler took personal control of Germany's armed forces. German troops entered Austria and took forcible control of that country, then Hitler announced support for Sudetan Germans in Czechoslovakia, threatening that country. Trenches for protecting people from bombing were dug in London, children were evacuated to the country and reservists were called up. In September Neville Chamberlain flew to Munich for discussions with Hitler and returned with the peace agreement on "the bit of paper". Hitler took mass action to persecute more jews in Germany and Japan captured Nanking and Canton. In 1939 German troops marched into Czechoslovakia in contravention of the Munich agreement. Italy annexed Albania and Germany annexed Memel. Then in August Russia and Germany unexpectedly signed a nonaggression pact, followed immediately by Germany invading Poland. Britain and France gave Germany a 48 hour ultimatum to leave Poland, failing which there would be WAR. By July 1939 it seemed to us in Ceylon that war was inevitable. When the declaration finally came on 3rd. September it was greeted almost with relief. We now knew where we stood, at last, We still dreaded what the future would hold, but at least we were not to sit back anymore, like frightened rabbits, and let the dictators brutality rule our lives.

The Jungle Cock's Call.

Of all the sounds of the jungle if I had to make a choice,
I'd choose to hear that clarion clear, The Jungle Cock's
"Jock, George, Joyce."

How it cuts the air of a morning, with its roistering boisterous noise, It is heard all over, that arrogant call,the Jungle Cock's

"Jock, George, Joyce."

Ah! I picture that flashing beauty and the jaunty martial poise of that swaggering knight, all poised for flight, as he challenges "Jock, George, Joyce".

Ceylon Jungle Cock. (Gallus Lafayetii)

Here in the dusty city, in my heart re-echoes that voice,
And I long for the day when I'll hie me away
Where the Jungle Cock calls "George Joyce".

                                        Gallinago

CHAPTER EIGHT

WAR

"The broad mass of a nation .... will easily fall victim to a big lie than to a small one"
Mein Kampf, Adolf Hitler
"It is the last territorial claim which I have to make in Europe"
(The Sudetenland - 1938) Adolf Hitler

"I have nothing to offer but blood, toil, tears and sweat"
Winston Churchill (13 May 1940)

At 2am on 4th September 1939 I was rudley woken by a bright light shining on my eyes and the roaring sound of engines. It was Gerald Du Pre Moore on his motor-bike, revving the engine outside my bedroom and shining his headlamp onto me in bed. He had ridden over from Nawalawatte to tell me that orders had come for us to report to No. 3 Company Headquarters at Kandy as soon as possible. War had been declared by Britain at noon Greenwich mean time the previous day, or about 5.30 pm Ceylon time.

We decided to have a good breakfast before setting off, but first we made a cocktail of all the liquor in my drinks cabinet and drank to our speedy victory and to the downfall of all dictators. Then Gerald returned to Nawalawatte to gather his kit, rifle and ammunition, and to get into his uniform. I did likewise and gave brief instructions as to the running of my divisions, for the care of Trickster and Pip, and contacted Jack Cranfield. He poor fellow would have the responsibility of being in single-handed charge of the whole property normally managed by a manager and four assistants. Jock Sutherland as second-in-command of the batallion was leaving for Colombo, and would shortly be taking over as commanding officer. Old M. Sinniah, the Head Kanakapillai on 2nd Division assured me not to worry; he would see that everything carried on just as if I was on the estate. By about six am I was off in the car to pick-up Gerald and head for Kandy. On arrival there, after a brief stop in Nuwara Eliya at the old Grosvenor Hotel, we were directed to The Queens' Hotel where everything seemed to be in confusion and nobody had any use for us. The obvious thing to do therefore was to bag a decent bedroom as quickly as possible. This turned out to be a wise move as late comers, and others who were slow to consider their position, had to bed down in a crowd on the Ball Room floor.

We must have stayed two or three nights or so at the Queen's and, so far as I can remember, very little happened apart from early morning parades in the street outside the hotel where our drill was watched by curious crowds. the hotel is situated in the centre of the town, There seemed to be plenty of time for relaxation. The tables in the Palm Court were all crowded with uniformed planters enjoying their beer or gins and tonic. His Majesty's Government was paying for our board; the future seemed uncertain to say the least, so why not eat drink and be merry before we went off to die? My open Wolsey Hornet was parked in Queen Street, outside the hotel and I did no expect ever to need it again, so it remained there when, on about the third or fourth day we were ordered to fall in with our arms and kit, and were marched to the railway station. We boarded a special train for Trincomalee, the great naval base on the East coast, and travelled through a hot and sticky night in hard-seated carriages to arrive at dawn. After de-training we marched through Trinco town, to the maidan, where tents were waiting to be pitched. The beach of Dutch Bay was only a couple of hundred yards from our camp, so after the sweaty work of pitching the twelve-man tents, and digging latrines, we all plunged into the sea for a refreshing swim. it was a very hot time of year with temperatures often well over 100 degrees F. That first night many of us left our kit and clothing lying on the bare ground in our tents and were horrified when, next morning, we picked up our bags to find the bottom, and much of the contents, had simply disappeared. The soles of one man's boots had been eaten. Termites had come up out of the earth and had a really good meal. I think it was Donald Marley's kit in our tent that had been selected as the most tasty.

For the next few days we were kept busy digging tactical defences between Back Bay and Yard Cove, which separates Trincomalee town, the Inner Harbour, and the Fortresses from the mainland. We all worked bare to the waist, the profuse perspiration which constantly covered our bodies ensured that we did not suffer from sunburn. Everyone was issued with a daily salt ration to replace the heavy loss of salt through sweating. When these inner defences were complete we were taken a few miles up the coast by lorry, to the ferry at Irrakkakandi, north of Nilaveli, to dig more defences. Next we practiced some simple manoevres designed to oppose Japanese landing parties. Not that our tiny force, backed by the guns of the Garrison Artillary on Ostenburg Ridge, could have been effective against anything other than a small raiding party. We were all waiting anxiously to see whether the Japanese would attack. There was virually nothing in their way.

After a while our platoon, under Lt. Hugh Meldrum, was moved from the tented camp to roofed quarters in Fort Frederick. The station H.Q. was situated in this fort and our platoon supplied the guard at the entrance gate. This ancient fortress was built by the Portuguese in about 1624, and the massive gateway was added by the Dutch and bears the date 1676. I found living in Fort Frederick very pleasant after the heat of the tents on the open maidan. The name was given by the British in 1803, in honour of Prince Frederick, Duke of York. A small herd of spotted deer roamed within the walls, and the grounds were shaded by banyan and ingasaman trees which spread over the "kachcheri" (former Government Secretariat), and Wellington House, both taken over the by military. The latter was named after the Iron Duke who convalesced in this house in 1799 after ontracting fever during the campaigns in India. It is said that he had been due to take ship for Egypt, to a new command, in the "Susannah" which went down with all hands in the Gulf of Aden, and that his stay in Fort Frederick saved him from death, and for the historic battle with Napolean at Waterloo. The fort sits on a rocky promontory accessible only by a narrow ridge separating Dutch Bay from Back Bay, with the Maidan and our camp on the landward side. On the highest and eastern most point, known as Swami Rock, stands a famous Shivite shrine known as "Tiru Kona Mallai (mountain sacred to Shiva), from which Trincomalee took its name.

We enjoyed our stay in Fort Frederick. The Guard Room,in the immensely thick walls adjoining the entrance gate, was relatively cool, and so was the barrack situated high up overlooking the sea and open to all the breezes. A few of us used to climb down the perpendicular rock cliff and plunge into the apparently bottomless sea. We were warned of the danger of sharks which were said to be attracted by the sacrificial goats cast off the Swami Rock close by, but I don't ever remember seeing any. Thanks to Hugh Meldrum (platoon commander), and to GortonCoombe (platoon sergeant), we were excused some of the more menial tasks which fell to the lot of the others in the tented camp. Gorton-Coombe was a huge man 6ft. 6ins tall, and had recently succeeded to the charge of his father's estate at Poonagalla. Actually we had more than our fair share of very large men in the platoon. Edward Wallace-Tarry, Charlie Strachan, Tony Kelly, Mike Michels, and Tony Wilson were some of those I remember who, besides myself, were well over six footers. I think some visitors to Station H.Q. felt rather intimidated. One of our tasks was to guard prisoners in the cells alongside our guard room. One such was a little cockney gunner of the Garrison Artillery who was serving a severe sentence for calling a Sinhalese officer of the C.L.I. a "dirty black bastard". He was a difficult customer and made several bids to escape whilst being escorted on exercise.

Towards the end of September I had to go to Badulla for some court case connected with some labourers quarrels on Spring Valley. After getting the necessary leave I took a hire car and was driven, by way of the seven coastal ferries, down the East Coast to Chenkaladi, and thence inland to Spring Valley. The ferries were manually operated by poles, oars, and sometimes by pulling on a wire stretched across the water from bank to bank. The width varied from a few hundred yards to a mile or more. Goats, cattle, bullock carts, a bus and a variety of good country peasants would be accommodated. Sometimes there would be a wait of an hour or more due to the ferry having just left for the far bank. If you were not in a hurry, and I was not, it was a most peaceful and relaxing way to travel. Everything on Spring Valley seemed to be going on more or less normally and after attending the court and staying a night with friends I took the train to Kandy where I found my car, just as I had left it; sitting outside the Queen's Hotel. I handed the car over to Walkers garage for disposal and continued back to Trinco. Here I learned that we were all to be sent back to our estates on leave. The words "on leave" were strongly emphasised, and we were told we were liable to be called back to our war stations at any time, and we were forbidden to leave the island without permission from The Manpower Board. It was emphasized that there was every likelihood that Japan might enter the war against us, and attack our Eastern Colonies. For this reason the British Government wanted to keep us all out East. We were said to be potential officer material, we had a good knowledge of the country and understood the people. Furthermore Britain was determined not to repeat the great mistake of 1914/15 when the flower of British youth was uselessly slaughtered in the trenches as cannon fodder. So they said that the best place for us was on our estates performing useful work at keeping the tea industry going and supplying the Empire with its favourite refreshment. Meantime we would all regularly be called up for periods of training.

The introduction of petrol rationing made life on the estate rather difficult and lonely. I bought a Ford V-8 touring car, a large six seater, getting it cheaply for Rs.450 (about £36), as it was a petrol guzzler. Then I heard that old Oswin Wickwar, late Surveyor General who had retired to Aislaby Estate, had an old 1919 Velocette motor-cycle jacked up in his garage. I went to see him and he insisted on giving it to me. It was in first class order and scarcely used. I fitted electric light and altered the shape of the handle-bars and soon had it in regular use. The engine oil was manually pumped and it had a most unusual speedometer called a Bonnisken Isochromus. After that I only used the V-8 when I could pick up at least three passengers with petrol coupons.

This was the period of the "Phoney War" in Europe, but the war at sea went on relentlessly, with the dangers from acoustic mines, submarine warfare and attacks on shipping by fast and well armed raiders. Ian Fletcher had gone Home on leave in about July,being relieved at Nawalawatte by Gerald Du Pre Moore, who had handed 1st Division over to E de C Norrish. The latter was the son of an army colonel, had failed Sandhurst,and had served in the ranks before taking up planting. Being on the reserve he was called to his regiment in India ju;t before hostilities began. He was killed fairly early on, during the Abyssinian campaign I believe. Ian Fletcher was recalled from leave, sailing in the "Britannic" with several other planters and some of their families. He left his new bride behind in.England. The ship was intercepted in the South Atlantic by a German raider, probably a "pocket battleship". The women, children;and as many men as possible were put into the lifeboats, but there was insufficient room for everyone. As the boats pulled away Ian and a few others were seen leaning nonchalently on the ships rail,then the raider opened fire with her big guns and sank the "Britannic". The boats kept together at sea for many many days and, after much suffering were sighted close to the South American coast many hundreds of miles from the sinking. They were picked up and taken to Beunos Aires. One of the survivors was D.P.Macdonald of Cannavarella Estate adjoining May Mallay. He arrived back in due course and told us all about his hardships,adventures and sometimes hilarious interludes. He was fortunate in being able to laugh at some of their predicaments. Felix and Dollie Fowler had also been on Home leave that summer and returned to Ceylon in November 1939 in a Japanese passenger liner, still neutral, in name at least.

News from Home made me envious and home-sick. John, my eldest brother was serving as a Lieutenant R.N.R. at Stubbington, then connected with the adjoining Naval Air Station at Lee. Admiral Bell Davis V.C. was in charge. Louis, brother number two, was already in France with the British Expeditionary Force, as a platoon commander in the Kings Shropshire Light Infantry, No.3., Tup (Cuthbert), was a regular with the other battalion of K.S.L.I, stationed in Bermuda. Francis was still with the Blue Funnel line and had just left Yokohama homeward bound when war was declared. I bet his old man (captain) was anxious at the thought of the Japs joining in. As it was the Japanese authorities broadcast to all the world the details of the ship, her gun and ammunition etc. This result-ed in a formal complaint of a breach of neutrality. On arriving Home Francis was appointed Midshipman in the Armed Merchant Cruiser "Derbyshire". This was the beginning of a war-long distinguished naval career. Andrea, first sister and No.6 in line, was doing her training at the Ipswich Hospital, and Margaret, No 7, was at the William Treloar Hospital at Alton. Guy, 8th and last,was still at school. My father was doing his damndest to get a war job, or a job which would enable someone younger to go on active service. However at his age, the mid sixties, he was unlikely to have much success. He was spurred on however by the fact that his old friend Captain H.H.W.Hughes RN., Retired, had managed to get himself appointed in command of an armed mer-chant cruiser. However the appointment proved too much for him and he had a heart attack whilst at Scapa Flow at the time of the sinking of the battle-ship "Royal Oak". He recovered and was, to his great embarrassment, appointed in charge of an establishment of WRNS. He was a life-long bachelor and rather scared of females. There were also twin maiden Aunts who had served as members of the Queen Alexandria's Imperial Nursing Service in the war of 1914 to 18, and.who had volunteered to serve again. They were attached to the 6th General Hospital, B.E.F., and would soon be off to France.

There was a general mood of optimism, or was it bravado, as was evident from the letters people sent each other at that time. On 11th October 1939, brother Tup wrote from Bermuda;-

"I wonder if you have had any scares - yet ? We had one the other night. Some sentry with an imagination saw a shooting star, or a flash of lightning, and sounded the alarm. It spoilt my Saturday night. I have planned myself a rather good war. Let's hope it will come off. I want it to turn out this way. stay here until January, then have a month's training at Home; get to France in time for the Spring Offensive; spend the spring and summer travelling steadily North East through Germany, and find myself on my birthday (Sept 2nd) sitting in a cafe in Berlin with a huge tankard of iced lager, my feet up on the table and Ein gnadiges fraulein in each arm Jawohl ! Strafe Hitler ! Deutschland.UNTER Alles ! God Save The King!"

Brother Louis had written a few days earlier to say; "I hear you were hauled out of bed at 2am to settle your account with Hitler! I'm simply furious with him and will make him sorry. Of course I am not allowed to put my address......."

It was not very long before I was back in Trinco for another spell of duty. This time we made our way independently and I gave Philip Grimwood, Paul Pern and Dennis Johnston. a lift in the V-8. Whilst at Trinco the "Queen Mary" and "Queen Elizabeth" were escorted into harbour by the Royal Navy. They were crammed with Australian and New Zealand troops on their way to the Middle East. Some of us got hold of a harbour launch and set off to view the liners which had only just been converted for trooping. We were forbidden to go close to the ships but we got near enough to see thousands of khaki clad figures; some with heads sticking out of portholes, some jammed together at the ships rails, and others clambering on any vantage point they could find. They were calling out "Is this Mombasa"? "Are we at Dar-es-Salaam"? "Where the hell are we?" etc etc.

In the evening the convoy sailed out of the harbour preceded by the aircraft carrier H.M.S. "Hermes" whose Swordfish torpedo bombers could be seen landing on board, presumably after having searched the approaches for hostile submarines. We had been allowed to bring one servant per twelve man tent for this camp, and my servant Sudubanda (=white man) had just arrived by train for his spell of duty. He had never been more than a few miles from his village before, being a simple Sinhalese villager, and had never seen the sea. Asked what he thought about it all he said the sea was wonderful, specially its salty taste, but there was nothing very special about the ships. After all they were not much different from a tea factory and there were plenty of those about.

This spell of duty in Trinco passed quickly and we returned to our estates once more. Later on we were to do guard duties and training in Colombo. All this made a change from work on the estate which was becoming rather trying owing to the fractious moods of the labour force. The workers had been having an easy time whilst we were away, and when the Tamil labourer is idle he is inclined to intrigue against his fellows or against his employer. This was compounded by trouble makers entering the estates, while we were away, and stirring up trouble, often using legitimate grievances which were outside the control of planters. Had we been present these could have been explained. The first strike ever recorded on a Ceylon Tea Estate took place during this period on Kotiyagalla, an estate noted for the stability of its work force, and was followed by strikes on a number of properties. Sri Pandit Nehru had visited Ceylon in July and at several mass meetings of Indian nationals had urged Indians to form an Association to protect their interests. This resulted in the formation of the Ceylon Indian Congress (now the Ceylon Workers Congress), and there is little doubt that this organisation did much to stir up trouble.

There was nothing on Spring Valley on which you.could put a finger, but there was a noticeable air of tension. This was the first time that I could sense, or feel in my bones, that trouble was coming. Later on, particularly during my last few years in Ceylon, I learned to "feel" a sense of serious unrest - whether on the estate or in the towns and villages -even when there was no visible or audible evidence to support any such feeling. Even though Spring Valley seemed quiet enough there was serious trouble on Roeberry and, nearer at hand on Demodera and Gowerakelle. Mass meetings of workers were held in Badulla with calls for island wide strikes. I attended one of these meetings with a fellow named Tony Wilson from Oetambe Division of Demodera. Tony was very fluent in Tamil. Just as we were about to set off for Badulla his Cook Appu asked for a lift into the town to do some shopping. As he got into the car Tony grabbed his shirt collar and gave a tug which revealed that the fellow was wearing no less than six of his master's shirts. Presumably he was taking them to sell or to give to friends and relations. As a result he was left behind and told to wash the shirts and return them to their normal place in the bungalow. I don't remember much about the meeting except that it was (fortunately) dark and we kept well in the background. There was a lot of noise but as the police were out in strength nothing very much was decided.

A few days later, after the police had opened fire on some unruly crowds connected with the Roeberry strike, there was a meeting between the Government Agent.of Uva and senior members of the Planters Association and the Police. I think the P.A. was pressing for the plantation industry to be declared an Essential Service. The view was that the disturbances were of a political nature and not connected with any particular grievances against estate managements.

As became normal practice later on, a long list of demands was submitted to various estates. I forget the precise nature of these but there was no doubt that the main cause for complaint was the sudden ban,imposed by the Indian Government,on all travel by estate workers between India and Ceylon. The notification of this ban had come without warning in August 1939, and it had resulted in many families being separated. Wives were separated from husbands, children from parents, and so on. Over the years the ebb and flow of labour between the two countries had followed the natural law of supply and demand, but the fact that the workers employed on Ceylon estates were able to return to India and, during such absence, retain a lien on their employment in Ceylon for an indefinite period, was unique in the history of labour relationship. When labour supply became ample, and work was no longer available to all-comers, the return flow was reduced by limiting the validity period of passes (Identification Certificates), and only holders who had been away for twelve months or less were given assistance to return. The fact remained that, within such a period, they were still entirely free to visit India and return to work at their own free will and pleasure, and at no cost to themselves. This measure had been understood and accepted by the workers for several years.

It so happened that the ban was introduced at a time when the seasonal flow to India had just passed its peak, and it was estimated that some 50,000 labourers who had gone to India were unable to return. So it can well be imagined what distress was caused by the new total ban imposed (apparently) unilaterally by the Government of India on its own citizens. The Planters Association and the Ceylon Government sent representatives to Delhi in efforts to get immediate alleviation of the distress, and although some cases of special hardship received individual sympathy, it was not until well into 1942 that the ban was lifted.

We were not accustomed to this sort of labour unrest. Nor were we organised to deal with it effectively. It all devolved on the individual planter to keep things going as best he could. In fact it took years for the employers to get their act together. It was not until about the end of the 1950s that we became capable of conducting negotiation effectively in a professional way. The Ceylon Trades Union had help from Russia, the Communist countries of Eastern Europe, Communist China, and from Left Wing Union leaders in Britain. Eventually the newly formed Ceylon Estates Employers Federation was able to give planters some excellent advice on individual and general cases of dispute. Well qualified Labour Relation Officers (LRO'S) were appointed to each planting district. They in their turn were supported by panels consisting of senior planters.

All this bad yet to come. Fortunately in 1939 and 1940 the majority of our people soon realised that the agitators were out to feather their own nests and were not in the least interested in the labourers welfare. So the trouble died down but did not go away entirely.

We juniors had little contact with officials such as The Government Agent (The Chief Provincial Administrator),The Police Superintendent, The Provincial Engineer, P.W.D (Public Works Dept), etc. They were all busy senior people with very little spare time who, on social occasions moved in a circle made up of some senior planters and Sinhalese landowners. However a new "Kachcheri Pup" came to Badulla and took up his residence in a bungalow close to the Kachcheri. The Kachcheri is the seat of a provincial government and, in the case of Uva, was housed in imposing office buildings on the site of the old star fort in Badulla. Some of the earthworks of this fort still remained. Other pretentious buildings nearby were the District Courthouse, the Government Hospital and the Police Station. The anglican church of St. Marks was also just across the green. All these buildings, and the official residencies, were attractively designed (often in the Kandyan style of architecture), and neatly arranged, with many fine flowering and shade trees forming cool avenues between them. It is a pity that the necessary additions in later years, to cope with the expanded needs of the Province, were crammed in between these fine properties in a makeshift and unsightly way, giving the impression of having been erected in great haste without thought of planning, not to mention the loss of so many beautiful trees. This sort of official vandalism has not been confined to places like Badulla and seems to be happening world wide.

"Kachcheri Pup" was the term used to describe a trainee civil servant learning the ropes and destined for an appointment as Assistant Government Agent,or other such responsible post,in the Ceylon Civil Service. The new arrival who soon became popular with some of us young bachelors, was V. Coomaraswamy who had recently returned to the island after completing his education at a university in England. It was Dennis Johnston of Mahapagalla, Ury, who got to know him best. Dennis introduced Noel DysonRooke, also of Ury, and myself, and a few others to "Coom". We found him to be very amusing and well able to laugh at us and at his compatriots. He would greet us with "Here come the bloody imperial white conqurors wanting to exploit my native whisky" To which we would reply "Of course, and some of your hottest vegetable curry you.lilly-livered black devil" then we would all burst into roars of laughter and slap each other on the back. Such behaviour would horrify today's advocates of "Correct Speak" whose ridiculous efforts do much harm to racial relations and to their cause for "sexual equality". Surely women are much better at some thing and men are better at others and nature never intended us to be similar? Do they not realise how America and Western Europe are derided for these misguided practices? "Coom" was quite open about his vision for his country's future and had he been alive would be appalled at events in Sri Lanka. I think we enjoyed his company so much as a refreshing change from the rathar insular attitudes at the club. "Coom" rose rapidly in the civil service to become Sir Velupillai Coomaraswamy KCMG.

There was not very much contact in those days between the European planters and the local landed and professional classes. The lawyers and doctors usually attended their own club, but there were inter-club tennis and cricket matches where we were able to mingle. We generally mixed very well on those occasions but their women folk tended to keep very much in the background. There were notable exceptions among whom were Dr. and Mrs, Thiagarajah who were extremely popular with the planting community. Later they moved to Colombo and set up in private practice becoming well known amongst the international set. I was always fascinated by the beautiful Mrs Thia who invariably wore a large gem stone dangling under her nose.

There was one aspect of communal life which struck me as rather strange. When Europeans gave a party to which Ceylonese were invited the latter seemed to be ignored by their western hosts and hostesses, being left more-or-less to their own devices. This struck me as inexcusably bad manners. It was very different when Ceylonese included Europeans among their guests. The latter were always given the impression that they were very welcome. Actually, as an European, I had received the same treatment from my fellows when I paid my first visits to the club and had assumed that, as a junior newcomer, I was beyond the pale. I have since noticed a similar attitude amongst some people in southern Britain. I wonder if the English are too reserved to take the initiative to welcome strangers, and only accept people after becoming accustomed to seeing them around regularly ? Whatever the cause I do not think rudeness was intended. After all in thls part of East Anglia it used to be said of the villagers that when a newcomer arrived "for ten years they looks you up, and for another ten years they looks you down;Then they accepts you".

Romance suddenly blossomed, or so it seemed at the time. Hugh Breay of Telbedde married Myrtle Sutherland's younger sister, Betty, who had come out to the island to visit her sister and brother Hugh Smythe, also an Uva planter. Soon after that,Dopey announced that an old family friend would arrive from Ireland with a companion. Kay Considine was petite, dark in a celtic way, and had the celtic fire. She was full of go and enthusiasm. Her companion, whose name I don't remember, was also charming, being tall, languid and blonde. It wasn't long before Gerald and Kay were married and the companion departed, but not before I had taken her around in my sidecar. I had disposed of my Ford V-8 when on a military course in Colombo, due to its excessive thirst for fuel, and had acquired a sidecar for the 1919 Velocette. A sidecar does not offer much opportunity for romance!

When attending CPRC courses in Colombo we were housed in Echelon Barracks which were within easy walking distance of the Colombo Fort, the shopping and business section of the city. We did weapon training on Galle Face Green and field training on the links of the Royal Colombo Golf Club. We also did our full share of guard duties, and provided guides for Australian troops sent ashore for route marches etc., whilst calling at Colombo on their way to or from the Middle East. The New Zealand troops were always charming and well behaved, but we had some troublesome customers amongst the Australians. We were called out to control some of the latter who had gone into Simes (a shop in The Fort), and had helped themselves to bottles of liquor off the counters, knocked the tops off the bottles and swigged the contents. They then grabbed some of the attractive Burgher (of Dutch descent) girl shop assistants. Another group had tried to get into the G.O.H. (Grand Oriental Hotel) to loot the bar, but had found the iron gates shut against them, so tried to break it down. We were called out and rushed to the scene to man the hoses, intending to drench the rioting troops. Unfortunately the hoses only produced a tiny trickle of water. However we some-how managed to shepherd all the troublemakers to the jetty and onto a trooper, where I believe they were effectively dealt with. This lot were on their way back to Australia in disgrace, and really should not have been allowed ashore unescorted.

When on duty at the Quarter Guard, the sentries had to stand in the full glare of the tropical sun during part of the day, and a puddle would sometimes form at our feet due to sweat pouring down our bodies and legs. Echelon Barracks was infested with lice. To discourage them climbing onto us at night to take their fill of our blood we used to place the legs of our hessian covered, wooden camp beds in tins of kerosene, but they were canny little brutes,and soon found that they could get at us by climbing the walls, crossing the cieling,and climbing down our mosquito nets. These nets were supported by ropes attached to pulleys screwed into the cielings. We went to work at fumigating the place and had it just about clear of the vermin by the time we departed. Whoever occupied the barracks before us must have been a lousy lot. During the south west monsoon the salt sea air blew right through the barracks, which kept us reasonably cool,if damp. The barrels of our rifles became rusty overnight even though left "bright clean and slightly oiled" the evening before. There was a good NAAFI with very cheap but excellent quality Nuwara Eliya beer on sale very cheaply. We drank gallons to replace the loss by sweating.

Although we were allowed occasional late passes up to llpm, we reckoned this was not late enough and soon found a way in and out of the barrack area without disturbing the night sentries. On one occasion I had been to a dance at the Galle Face Hotel where I had wined and dined very well. I found Noel Dyson~Rooke there dining with his parents, and was invited to join them. After dinner Noel went back to barracks but I stayed on in the hotel bar. Either I imbibed too freely or, as sometimes happened during the war, the drinks were doctored. I still had the V-8 then and was trying, in the early hours, to drive it across the Galle Face Green road which was deserted at that hour. Suddenly I collided with the one and only other car, which was coming in the opposite direction, i.e. towards the Hotel. This quickly sobered me up and I got out to see how things were with the opposition. However the other car was too quick for me. It backed away, and then passed me with much scraping and grinding of metal in pain. I noted that there was a man and a woman, both in evening dress, in the car. I turned back to follow them to the hotel but when I got there they had disappeared. My car was not going too well, so I left it outside, down by the shore, and walked back to barracks, arriving in time for Reveille. My car was in the same place, untouched, when I managed to get out of barracks a few days later. I heard nothing from the other party. I kept my eyes and ears open but there was never any mention of the accident. They must have had a good reason to keep quiet.

My recollections of the order of events at this time is rather muddled and I may have got some of the sequences wrong. I was usually glad to get back to the estate after military duties, but there was always a lot of catching up to do, and with others away life became very lonely. However some of us still managed to get away for brief week-end shooting trips. These were absoltitely terriffic fun. I was a poor shot but that did not detract from the marvellous atmosphere of dawn and evening snipe and duck shoots in wonderful country. A contingent of 50 CPRC members was selected and sent to the Middle East. The Manpower Board controlled the selection and my name was not included.

I was on the estate when the "phoney war" came to an end with the Germans invading Norway and Denmark on 9th April 1940. By the 15th of that month British troops had landed to oppose them and great naval and military actions took place at Narvik,' and along the coast, but it was too late. I spent several months of nail biting, stomach churning anxiety, and hours of listening to the news in my lonely bungalow. Disaster followed disaster until the miraculous evacuation took place from Dunkirk and Calais. It seemed,after that, that only the Royal Navy and the tiny force of RAF fighters stood between Britain and an allpowerful Germany which had every prospect of pounding my country into submission.

British forces were quickly withdrawn from Norway, loosing many men and ships in the process.Holland and Belgium were invaded on 10th May and Holland ceased fighting on 14th. On 27th of the same month Belgium capitulated. Ostend, Ypres, Lille and other French towns were lost to the German's. The roads were clogged by refugees who were mercillesly machine-gunned by the Luftwaffe. Then between May 27th and June 4th, 299 British warships and 420 other vessels, under constant attack, evacuated 335,490 officers and men through Dunkirk. Hitler next proclaimed a war of total annihilation against his enemies and German forces penetrated French defences near Rouen. On 10th June Italy declared war against Britain and France and Paris was captured. On 15th Russia occupied Lithuania, Latvia and Estonia, and on 25th hostilities ceased in France. In July Germany occupied the Channel Islands, Japan invaded Indo-China and stepped up her attacks in China. A squadron of the Royal Navy immobilised a French fleet in Oran to prevent it falling into German hands. Britain and the Commonwealth stood alone against Hitler's hordes which now controlled all Europe. Italy was now openly against us and it looked as if Japan would penetrate further west any moment.

Later in 1940 another contingent of CPRC men left for the Officers Training Unit at Bangalore in India. Once more I was unable to get my release by the Manpower Board. Several good friends got away including Ian Mackenzie and Noel Dyson-Rooke. In due course these two were commissioned and posted into the same battalion of the Jat Regiment. The unit was sent to Malaya and was right up near the Siamese border when the Japanese attack started.

Letters from Home were taking two or three months, or even more, to reach us. News filtered through that Jack Wallace, from our sister firm, the Ouvah Company, had been taken prisoner outside Calais. He and his wife Marion were on Home Leave when the war began and, being an ex regular officer, he was on the Reserve so reported for duty in Britain. My brother Louis had been safely evacuated with his men, through Dunkirk after many adventures during the exhausting retreat through Belgium and Northern France. He reported that the retreat was perfectly orderly and whenever they managed to come face-to-face with the enemy he and his men got the better of the exchanges. He married Jill Spencer during the short leave which followed his arrival Home. Twin aunts ,Hilda and Eve, also came Home safely via Cherbourg after many close shaves, retreating in their hospital train crowded with wounded soldiers. One evening not very long afterwards they came home from hospital to find their house in Norwood had received a direct hit during an air raid. There was little left of their lifelong possessions that could be salvaged. Francis, a younger brother, had volunteered for the Royal Navy's Bomb Disposal Unit. He was based at the Dorchester Hotel, in Park Lane, and had a special car and chauffeur standing by 24 hours a day ready to rush him to places like Coventry, Manchester, Liverpool, Sheffield or anywhere else that reported German delayed action parachute mines being dropped. He was entitled to cross all traffic lights at red, and to travel the wrong way down one way streets. My father was doing duty with the Home Guard, ready to do battle with any of Hitler's men who might land near the Suffolk coast.

In September I wrote Home saying "On Tuesday a coolie brought me a young crocodile two feet long. He lives in an old glass acid tank and snaps and snarls if you put a pencil near his mouth. In fact he isn't at all friendly. I am going to turn him loose as soon as I can get down to the Badulla Oya (river)...... I should, I suppose, be thankful to be in such a peaceful spot, but that is not possible in the present circumstances..... It seems all wrong that blokes as young as some of us out here should not be allowed to go (off to the war) whilst at Home middle-aged men, married with families, are being conscripted to'fight".

In December I wrote to Colonel Sutherland as follows:-
"I should like to thank you for the efforts you have made on my behalf to obtain the Directors permission to allow me to apply to the military authorities to leave the island with a CPRC contingent. I have very much appreciated your efforts in this direction.
I should also like to make it clear that I am quite prepared to resign my billet here should there be any possibility of my being accepted for any contingent, or for any post in the army or navy, without my employers sanction. I should appreciate it very much if you would inform the Directors of my intention."

The reaction to that letter was negative. I was referred to the Manpower Board who, the Directors said, decided all such matters. So life continued much as before with two weeks away on military training every six weeks or so. on the estate I missed the bachelor companionship and high jinks with Dopey. Somehow it wasn't the same any longer with Kay around even though she couldn't have been any kinder or more considerate. Norrish's place had been taken by another married man, Charlesworth, so I was in a minority of one bachelor to four married men. The 5th Division kanganies Arunasalem and Vemban continued to give me first class help and support, as did old Sinniah Head K.P. (kanakapillai) on second Division. The N.E. Monsoon through November 1940 until February 1941 was a very wet and prolonged one, so up in the continuous mists and drizzle on May Mallay it was depressing. My recollections are of the damp smoky smell of wet "cumblies" (waterproof sheep or goat wool rugs worn over the heads of field workers) which were dried out overnight over wood fires, the scent of wet green leaf in the pluckers baskets as they lined up for the evening weighings, and the drip, drip, drip off the trees round the bungalow which went on day and night.

Then the Du Pre Moores came to the conclusion that I, being senior to Gerald, was blocking his promotion; which I probably was. So they started looking around for another billet with better prospects. I felt that if they succeeded in getting away my prospects would be even worse. With Gerald's very poor eyesight there was little hope of him getting any active service appointment. However he did get another planting appointment and I had to run his 3rd and 4th Divisions in addition to my own.

In March 1941 I wrote home saying;-

"It is difficult to get leave with so many military camps and with one or other of us (from Spring Valley) attending them. Mails are still taking a long time in coming. Last week I received a batch of Christmas cards and your letters of December 4th and 30th. They were simply full of news. Thank you so Much"

On September 23rd. 1941, my father wrote;-

"A, merry Christmas and a 'Happy New Year and much love from us all at Home..... I hope you will get this in time for Christmas".

The main news was about my younger brother Francis being awarded the George Cross. Actually I had already heard about it in June from Flabby Fowler and Saibo Sim, who had picked it up on a BBC news report. They had both telephoned me immediately to let me know the great news. In all, during his bomb disposal work, Francis dealt with twenty unexploded, delayed action parachute mines. Some had fallen in almost inaccessible places; e.g. one was suspended from its parachute which was caught up in the roof over Sheffield's main railway station. Another had fallen inside a deck locker attached to the engine-room casing of a ship in the Manchester ship canal. He had to deal with this by inserting himself, head downward, into the casing. He could not see the timing device but could hear the ticking of the clock on the bomb fuse. He somehow managed to place a "gag" on it by feel. The citation stated;

"In spite of this Sub-Lieutenant Brooke-Smith remained in position pressing down the gag and finally managed to secure it properly. There was no doubt that this cool and courageous action prevented the mine detonating. It is mentioned that this is the first time that Sub-Lieut. Brooke-Smith had used a safety gag on a bomb fuse, and he could not see what he was doing with it. At the best his chances of escape from this mine were very doubtful".
Twenty mines were enough,and his next appointment was to H.M.S. "Broadwater", an old ex U.S.S. Naval destroyer handed to Britain as part of the "Lease Lend" agreement. He was immediately sent off to escort convoys across the North Atlantic. One of his letters mentioned "dodging icebergs"

which gave a clue as to his whereabouts. It could scarcely be said that he had jumped from the frying pan into the fire.

The rest of the family were all "doing their bit". Eldest brother John had been appointed to H.M.S. "Conway", the well known school training ship stationed off Rock Ferry in the river Mersey. Louis was in Lincolnshire training his troops, Tup was back from the West Indies doing "special" training in Dumfrieshire. He managed to send the occasional brace of pheasants home to our parents. The girls were nursing or with the WRNS and the youngest, Guy, was "settling in well at Framlingham College". The indomitable twin maiden aunts were still enjoying their adventures. They were nursing wounded German prisoners of war in the hospital ship "Dinard", somewhere on the coast, and expecting to set sail with them to Germany at any moment, to exchange them for wounded British prisoners. The operation was eventually called off at the last moment and they were transferred to Netley Hospital.

During that summer (1941) I was not too fit, so I tried applying for Home Leave on grounds of health. I also pointed out that it was nearly six years since I had left Home and, by normal conditions leave was long overdue. I felt I was losing touch and was beginning to wonder whether I would ever see any of my family again. My plea to be allowed Home was met with the offer to issue me with a permit to go to India for one month's leave. My inclination was to tell the authorities to go to Hell. However after discussion with Sutherland and Cranfield I took their advice to accept the leave offer, and started to make arrangements for a trip to the Nilgiri Hills in South India, intending to go on to Bangalore and return through Travancore on the West Coast.

I left for Colombo by the night mail on 14th October, arriving at Fort station at 6.30 the following morning, intending to collect my papers and catch the night mail for India the same day, in the evening. However nobody at the emigration offices seemed to know anything about my case and I had to stay the night at the G.O.H. The following morning I called at the office of the O.C.Troops Ceylon, then a Colonel,who signed my permit on the spot. So I caught the night mail to Talaimannar just 24 hours late. A Mr. and Mrs and Miss Arden, from Madulsima, were on the train, going to Poona. There were also two CPRC officers on their way to a tactical course. We reached Talaimannar at dawn and boarded the ferry to cross the 20 mile Palk Straits to Dhanuskodi where we arrived to find the train of the Great South Indian Railway waiting on the pier, and an excellent breakfast ready in the dining car.

Owing to the delay in Colombo my berth reservation had expired and there was no place for me in the first class, so I travelled to Trichonopoly in a second class compartment, just managing to squeeze in with a crowd of Chetties (A caste which provides businessmen and "kadai karans"= shop keepers). The Indian trains had four classes of passenger accommodation, Ist, 2nd,Intermediate and 3rd. We travelled north all day across a dead flat dusty desert. Or that is what it seemed after the lush green of Ceylon. The approach to India's coast had been like going into Port Sudan, and Dhanuskodi consisted of nothing except the railway pier and a few fishermen's huts. At Trichy I changed trains and caught the 11.30 p.m for Mettapallian,, When I woke next morning we were still crossing a vast flat plain, but now it was very fertile with a rich soil supporting fields of paddy, maize and cotton. At Mettapallian I breakfasted at Spencer & Co.s station refreshment rooms before board-ing the narrow guage mountain railway. This took me up the escarpment to Coonoor at 6,000 feet above sea level, where the air felt deliciously cool after the stifling heat of the plains. At Clovelly, the Guest House where I had booked accommodation, I was greeted by "two funny little old things, the Misses Marjoriebanks, (pronounce Marsh Banks)," It had cottages, or cabins, in the grounds, and meals were taken in the main building. There were several service people staying there on leave, also a couple of "abandoned wives" whose husbands were in Malaya or the Middle East. My cabin was semi-detached. In the other half was a young "abandoned wife" of a Captain who was with his Indian regiment in Malaya. She was in a bit of a state as the Japs were being very belligerent. In fact it looked as if they would attack at any moment. The all-in charge at Clovelly was Rs.5/- a day, I wrote Home that "Coonoor is rather like some Surrey village such as Woodmanstern. I have joined the Wellington Club and the pro is giving me golf lessons. I found Mr. Zachariah's house and went in to find the family at tea and Miss MacArthur also there". The Zachariahs were Indian friends of my Missionary uncle in Bengal and I presume Miss MacArthur was also a missionary. I had introductions to several people, including Mrs. Willoughby-Grant, Major Dickinson at the Cordite factory, and to Mr. & Mrs Sisnett, but after a day or two I went down "with a bug". I saw quite a bit of the distressed abandoned wife and we played tennis at the club. Later she accused me of being inexperienced, but by that time I was far too ill to be interested. Probably just as well. I met another girl there who I described in a letter as an "exceptionally nice girl, the daughter of a judge in the I.C.S." I remember nothing about her. Someone took me to Ootacamund and to see some Todas the mysterious white race who live in the Nilgiris and practice polyandry. On 30th October I wrote to my parents as follows:-

"I had an awful shock when I read in yesterdays paper that H.M.S. "Broadwater" has been sunk in the Atlantic. There was no mention of survivors. However a Commander Hunt RIN., told me that she was probably sunk some weeks ago and that all next of kin have most likely already been informed. Since I have heard nothing from you I must take no news as good news and at the same time hope and pray that dear old Frink is safe and well. I have been having a grand time but unfortunately have, at the moment, a temperature and am in bed. Also I have made the disappointing discovery that I have still got my little friends inside me. I was supposed to have been cured before I came over here. I am g6ing to a doctor tomorrow if I can get up. I had arranged to play golf with an RAMC man named Menzies who is on leave here, and might come over to Ceylon with me for a bit if he can get a passport.

You will probably get this by January and by that time I hope the end of the war will be in sight, as I have no doubt that if Russia can hold the Germans up over the winter, we will launch an offensive in the West and North Africa in the spring. I hope the food question is easing and that all are safe and sound. I am desperatly anxious for news.

P.S. 2nd November; This morning a telegram arrived saying; 'ADVISE RETURN IMMEDIATELY REPORT H.Q. FOR INSTRUCTIONS SUTHERLAND". That means there is going to be a contingent, and here I am feeling lousy with 'flu and an upset tummy. However I am catching tonight's train, and should be in Colombo on Tuesday morning. I can't possibly miss this opportunity". My RAMC pal gave me some "dope" which he said should fix me for the journey.

Actually I had a foul journey. I felt so rotten that I broke it at Trichynopoly, spending the night in the station rest rooms, and visiting a doctor in the town. As soon as I reached Colombo I called at H.Q. to find there was no great hurry as the Man Power Board was delaying things. I might have guessed.

Back on Spring Valley once more I waited impatiently for news about my release and on 12th November wrote Home telling my parents that "if I am turned down this time I know it will be the

Agents fault and I will try to get an interview with the O.C. Troops or the Chief Secretary". A week later I wrote that "General Wavell has now sent one of his own men, Major General Inskip, to command all troops in Ceylon, so we are hoping that things will start to move at last. I have not yet heard from the Manpower Board, but am trying to get someone to look after Trickster and Pip and my motor bike. Saibo Sim was bitten by a rabid dog last week and was rushed to Colombo for the 14 very unpleasant innoculations at the Pasteur Institute. It must be very hard and tiring work for you Mum, taking the meat pies about, and collecting all the War Savings, as well as running the house for a family who come and go at unexpected times. You mention having written on August 5th, Dad, but I haven't had a letter for a very long time."

Then a few days later; "the first batch of 25 men - mostly Colombo wallahs - is going off to India early in December. Another batch is expected to go shortly afterwards. The Chief Secretary has written asking my employers to state their reasons for saying that I am indispensable!!. As Sutherland is backing me I think they will have to climb down. It is nice to know my firm thinks I am indispensable. I wonder if they will be in the same frame of mind when I want to come back after the war? Old Tup Grant-Cook is buying my Motor bike and sidecar. He now has to overlook many estates and, with petrol rationing getting tighter, finds it difficult in his big car. He is going to go in the sidecar and his driver will work the bike. I am going to store everything here and Sudubanda will be caretaker of the bungalow.

"Yesterday I went over to Glen Alpin to see John Benest. While I was there it pelted with a capital 'P'. It beat any artificial downpour produced by the film industry. He tried to make me stay the night but I prefer my own bed and left at 9pm. Coming up to 3rd. Division my headlamp bulb fused and I was left in pitch blackness. I found my way to some cooly lines and scared the life out of them by appearing like a Frankenstein monster, with my large mac and heavy canvas cape over it, and baggy oilskin trousers. After assuring them that I was not a "Pessassey (devil), I borrowed a hurricane lantern and walked the rest of the way, leaving the bike in a shed. The rain is still roaring down onto the "Tagram" (tin) roof, and shooting off into the concrete drains round the bungalow". 28 November 1941. "Woopee! London refused to allow me to go unless someone could be found to take my place and I was most frightfully fed up. Then old Colonel Maxwell Johnson, late of the Ceylon Mounted Rifles, heard about it. He retired ages ago and must be at least 70. Anyway he will act for me for the duration, so that I can get away. Isn't that grand? Sutherland has just cabled the information Home and I haven't the least doubt all will be OK. I think it is absolutely GRAND of the old 'Field Marshal' as he is called. He will put the fear of the devil into the coolies up here, but they love him wherever he goes. He went to Wewesse after the superintendent collapsed owing to riots and had ground glass put in his, and his wife's, food, in 1939 and had them all eating out of his hand within a month. He is taking on Trickster and Pip, and I hope will take Sudubanda too. Dennis Johnston is going with the first batch, and Mike Horsfall, Duncan Thompson, Tony Windus and Philip Grimwood and others have been told to report for medicals on Monday"

My father wrote on 7th November to give me news about Francis but, as has been seen, I had already heard about the loss of his ship. News of Francis' safety reached me by telegram when I got back to Colombo from the Nilgiris, but it was a long time before the full details reached me, as told to my father by the ship's doctor, Surg. Lieut. A.H.Clark RNVR of New Brunswick, Canada, and by Francis

"Broadwater" was one of the escort vessels to a convoy in the North Atlantic in mid October. The first two days were quiet, then on 17th October an enemy "U" boat was sighted on the surface. They gave chase but the "U", boat crash dived before they were close. However after dropping a pattern of depth charges an additional explosion was noted, and some of the crew reported seeing oil. Early next night there was a huge explosion and fire. An oil tanker had been torpedoed. "Broadwater" picked up some survivors. Three more ships went down before midnight, and from one only a single survivor was picked up. The doctor put him in his own bunk, and to that he owed his own life.

About; half way through the middle watch "Broadwater" was on the port quarter of the convoy with the captain asleep in his bridge cabin, and Francis asleep in a cabin also near the bridge. The "doc" was asleep in the captain's harbour cabin. All these were further aft than the main officers and mens accommodation. Francis came-to in complete darkness, hearing escaping steam and the absence of engine noise. He found he was suspended with the lower part of his legs over a bulkhead, into the next compartment, which had become separated from the deck above. His head and shoulders were covered in oil. He reckoned it took him about 15 minutes to find his way out, and in doing so he was able to notice that a gun had been blown into the cabin from the fore part of the ship. He eventually got to the bridge where he found the captain alive but very badly shocked, and the fore part of the ship had disappeared.

Clerk, the surgeon, also woke in utter darkness. After trying, for a Long time to find a way on deck, he began to lose- his head. Taking a grip of himself he decided to rest briefly and think things out. Having got a hold of himself he felt his way along until he came to a hole through which he passed into the sea. It was icy cold. After a time he saw that part of the ship seemed to be remaining afloat so, with great difficulty, he managed to get back on deck having to push two dead men out of the way. The climb up to the deck was a terrible effort. He found a steam pipe, still hot, and lay down on it to get some warmth. Other escort vessels sent their boats and picked survivors out of the sea, and off the decks of the floating portion. Some 60 officers and men were killed by the explosion or drowned, or went down in the fore part. All the people picked up by "Broadwater", from earlier sinkings, were also lost. The after part was sunk by gunfire when daylight came, and the survivors were put ashore at Londonderry on 20th October, but some of these subsequently died. One of the men lost was Lt. Parker RNVR. He was the first United States citizen to die serving with the Royal Navy during this war. Britain and the Commonwealth were still alone holding the Axis at bay.

But not. for much, longer. On 6th December the Japanese attacked Pearl Harbour and the United States entered the war, two days later the Japanese landed troops at Kota Baru on the North East Coast of Malaya. The dreaded moment we had all been waiting for in the East had arrived.

"Jonnie" Maxwell Johnson arrived up at May Mallay, with difficulty, on the last day of November. With difficulty because his car was an ancient 1920 Fiat and he ran it on coconut oil, or so he said. He had some coconut properties in the low-country and claimed that he refined enough to run his car. I had no reason to disbelieve him, especially as the car was very difficult to start, usually need-ing a gang of five or six men to push it. It also pulled very badly and a gang had to be sent to help push it up the steepest approaches to the bungalow. I attributed this to the poor quality, of his motor fuel. He soon had his kit unpacked. Most of it seemed to consist of old plate and china from the Queens Hotel, Kandy, which he had managed for a time after retiring from planting. His wife was living in Scotland. It took me a couple of days to pack my things, hand over, and say my farewells, then I was off to Colombo by train. Our contingent was encamped in Colombo for a few days during which we heard the terrible news of the, sinking of HMS "Repulse" and "Prince of Wales" off the East coast of Malaya, and that our troops were falling back on land. General Inskip gave us an inspiring talk, then after a wild party we boarded the train to take us North on our way to the Officers Training course at Belgaum.

Namunukula from Judges Hill.
Badulla, with Rock Hill in middle distance and Spring Valley on right.

PART 2

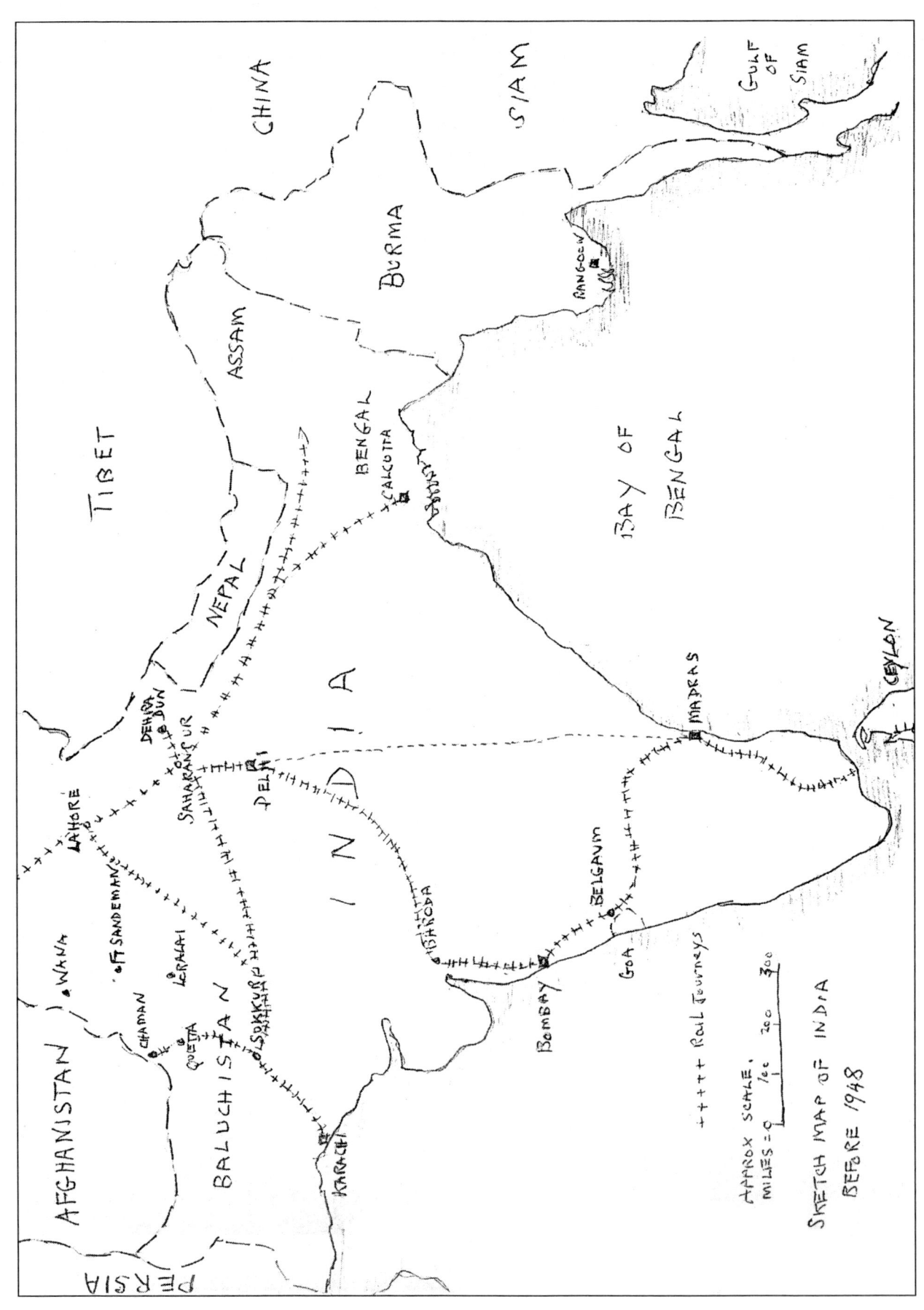

SKETCH MAP OF INDIA
BEFORE 1948

APPROX SCALE,
MILES = 0   100   200   300

+ + + + RAIL JOURNEYS

CHINA

SIAM

GULF OF SIAM

BURMA

ASSAM

RANGOON

TIBET

NEPAL

BENGAL

CALCUTTA

BAY OF BENGAL

AFGHANISTAN

WANA

FT SANDEMAN

CHAMAN

QUETTA

KALAT

LORALAI

BALUCHISTAN

SUKKUR

KARACHI

LAHORE

DEHRA DUN

SAHARANPUR

DELHI

I N D I A

BARODA

BOMBAY

GOA

BELGAUM

MADRAS

CEYLON

PERSIA

108

CHAPTER NINE.
BELGAUM - DEHRA DUN - BALUCHISTAN

"As a rule man's a fool
When its hot he wants it cool
When its cool he wants it hot
Always wanting what is not."

Edward Lear.

Our train puffed slowly out of Colombo Fort station that hot and sticky night, while sisters, wives and girl friends ran alongside to the end of the platform clutching hands., waving and calling out last words of farewell. As all those wellwishers faded from sight we withdrew to our compartments, mopping our sweaty brows, and started to prepare for the night's journey to the North. We had each been handed a parcel by a group of ladies who had come to see us off. These good Colombo people had probably spent many hours of literally sweated labour in making and parcelling up their gifts. We found that these contained a variety of hand knitted garments suited only to severe winter conditions in Arctic or Sub Arctic areas. As few, if any, of us could then forsee ourselves donning thick cap comforters or woollen gloves I doubt if any survived many miles of our journey without being flung from the carriage windows. Later at least one of us regretted casting them away so hastily.

Company Sergeant Major Reggie Marks, a stockbroker by profession, was in charge of the contingent for the journey to Belgaum where we were destined to spend three months on a crash course whose object was to turn us into Emergency commissioned officers (ECOs) in the Indian Army. I was fascinated by the country which we traversed, and the diversity of its people; all very different from Ceylon. The first part of the journey merely retraced the the already familiar section covered during my broken holiday of the past autunn. The first night took us to the Palk Straits separating Ceylon form India. After crossing the straits from Talaimannar to Dhanuskodi, another day and a night was spent in reaching Madras Egmore station. We had the best part of a day to explore Madras before taking another train at the Fort station* on the M.S.M. Railway. In Madras I, with a couple of others, even managed to penetrate the confines of the exclusive Madras Club where a local member, a friend of one of us, entertained us to a very good lunch. The MSM Railway train cut across the centre of the Deccan and we were glad to use the blankets provided to keep us warm during the two nights of the journey, for the night air was noticeably chilly compared with the accustomed balmy breezes of Ceylon. The train stopped for meals in the station restaurants at places with exotic names such as Guntakal, Bellary Hubli and Dharwar. At one of these, probably Hubli or

*Madras & South Maharatta Railway.

Dharwar, we changed from the Broad to a Metre guage line. At about the same time we came to the thickly forested and hilly Western Ghats which mark the edge of the central plateau before falling away to a narrow coastal plain bordering the Indian ocean. And so we came to Belgaum where we detrained and marched to The Officers Training School. Belgaum is situated within the Ghats, about 50 miles from the coast - as the crow flies-, and only about 10 miles from the boundary of the then Portuguese, and neutral,enclave of Goa. The military headquarters at Poona lay about 200 miles to the north. I had thoroughly enjoyed my first really long Indian train journey, and had soaked up all the sights, sounds and scenes which inevitably accompany rail travel in the subcontinent. 'The comparatively slow train journeys across the ever changing Indian countryside gave an impression of immense distances and variety; far more so than any journey half round the world in a jet plane. You saw the changes taking place as you travelled; met the different people at the stations and on the trains; heard the different languages from Tamil, Telegu,and Hindi to Bengali, Gujerati, Punjabi and Pushtu, and many more. There were the cries of "Chae wallah", "Hindu Pani" etc. etc. heard above the immense hubbub of thousands of voices as the train pulled into a station, the rumbling of the wheels silenced as it came to a standstill, and a quieter background of the electric fan whirring away at the stifling hot air in the cabin.

We found the previous CPRC contingent, mainly of Colombo wallahs, well settled in at the O.T.S. They greeted us with much helpful and kindly advice. We were issued with bicycles and rode in platoon formation from parade to parade and lecture to lecture. Dare I mention that one or two had to learn how to remain aboard and to control their metal steeds? Captain R.E.G.Twelvetrees (obviously he became known to us as "Pannyrendumarram" meaning "twelvetrees" in Tamil), was our Company Commander.He wore the uniform and badges of the 9th Gurkha Rifles.

We were kept so busy that we had little time,or inclination, to heed the news of daily disasters in Hong Kong, The Philippines, Malaya, The Dutch East Indies and Burma, where the Japanese were making lightning advances, destroying everything as they went like a prairie fire. It was all too easy for them after years of secret planning and preparation. The allies were badly handicapped in the East because events in Europe, near the hub of Empire, had meant that every available weapon was needed to defend Britain against the threat of invasion. Later when that threat receded the Germans were at the gates of Moscow, so Russian needs received priority and the Far East continued to be starved of men, weapons and materials, necessary to fight a ruthless and very efficient enemy in what could be called his part of the, world. It was not known until after the Japanese war-machine was set in motion, in December 1941, that they had spent years of careful preparation and planning for their campaign.

Our Ceylon contingent made up two platoons of A Company. The third platoon consisted mainly of men from Calcutta, Bombay and other Indian stations, plus a few men who had got away from Malaya and China. Another company was comprised of Junior Officers of the British Army sent out for attachment to the Indian Army. They were sent to O.T.S. Belgaum, to learn about the country, to speak Urdu, the "Lingua Franca" of the Indian Army, for acclimatisation and assessment prior to allocation to specific Indian Army units. We old Indian hands tended to look down on those fellows. Many of them seemed callow and although they had received their commissions in Britain, and already held rank as second lieutenant, or lieutenant, while we were yet only "cadets", their basic, training seemed to us to be sadly lacking in soldierly qualities, and their attitudes were sloppy. There were a few notable exceptions. The Ceylon contingents challenged all comers, including instructors and staff to cricket, Rugger, hockey and golf matches, and ended as the winners in all the sports. On the parade ground our CPRC training under senior NCOs of the Guards Brigade and the Rifle Brigade, gave us at least a head's start. The fact that we all knew each other very well also counted for a lot.

After an early reveille and physical training followed by a short break for breakfast, our days were occupied by a mixture of weapons training, square bashing, route marches, driving and maintenance of vehicles, Tactical Exercises Without Troops (TEWTS), as well as excercises with troops, demonstrations and lectures. The lectures were intended to familiarise us with the many and immensely varied units of the Indian Army, and the multitude of classes enlisted therein. To name a few of the hundreds of classes there were Dogras, Garhwalis, Kumaonis, Gurkhas, Gujars, Ahirs, Rajputs, Maharattas and Madrasis. The former were Hindus other than Sikhs and Jats who were also subdivisible into numerous classes. Then there were the P.Ms (Punjabi Musalmans), and other Mohammadans, ranging from those hailing from Bombay and Madras Presidencies to Musalmans from Rajputana and the Delhi Districts etc. We learned about the Shia and Sunni sects, and what units recruited which class. The Pathans, the tribesmen that inhabit the North West Frontier area, together with their history, characteristics, the type of life they lead and details and locations of the numerous tribes, were allocated an entire set of lectures, for historically the greatest threat to India had always been from that area.

During the evenings we spent an hour or so with our "munshis" (language teachers) , learning the intricacies of Urdu. We were required to pass the elementary Urdu examination soon after joining our units. Without that qualification promotion was barred in the majority of units, and rightly so. Thereafter some would continue to develop their Urdu whilst others would learn Punjabi, Gurkhali, Pushtu, or whatever language was generally used by their men.

A brief description of the, political situation in India during the years leading up to and early in the war, seems appropriate here. Some eminent historians were of the opinion that even though the vast majority of the population in the country were indifferent to politics, and had little or no interest in the machinations of either the Indian National Congress or the Muslim League, a number of educated townspeople had, since the beginning of the century, become embroiled in politics. M.K. Gandhi was regarded as a saint and hero, or as the devil incarnate, depending whether one was a Hindu, Muslim or Christian. Acts of anti government terrorism by extreme nationalists were centred almost entirely in Bengal, but periodic campaigns for civil disobedience took place in many parts of the sub continent.

During 1937 to '39 communal strife increased considerably with Hindus and Muslims becoming extremely suspicious of each others motives. During that period communal violence (mainly in Congress run provinces), resulted in over 2,000 casualties. A considerable degree of self government had been achieved and, by the end of 1941, the Viceroy's Executive Council had a majority of Indians for the first time (eight to five). However Congress dismissed this as a few "Yes Men" under the thumb of the Viceroy.

On the outbreak of war in September 1939 Gandhi had told the Viceroy, Lord Linlithgow, that he viewed the war with an. "English Heart", and was in favour of giving England and France unconditional support but, as he explained later, in view of his belief in nonviolence, this could only mean moral support. Nehru also admitted that in a struggle between democracy and fascism, Indians sympathies were inevitably on the side of democracy, and he would like "to throw all her resources into a struggle for a new order". He also said that Congress was "not out to bargain" and take advantage of Britains' difficulties.

The majority of educated Indians were more keenly alive to the iniquities of Hitler than they had been to those of the Kaiser, and several British officials who had been in India in 1914 thought there was a more genuine desire to play a part in Germany's defeat than there had been 25 years earlier. The Muslim League also condemned Nazi aggression, whilst the Princes assured the Viceroy of their full support. In May/june 1940 the successful German Blitzkreig, the entry of Italy into the war and the fall of France, suddenly transformed the war situation, making not only a British victory seem impossible but Britains' very survival uncertain. Yet Britains' extreme peril excited great sympathy in India. Nehru offered to throw Congress's full weight into the war effort. In return he asked that Britain should grant full independence to take effect at the end of the war, and that greatly increased power be granted immediately to Indians in the Legislative Assembly.

After the fall of France and the closure of the Mediterranean, encouragement was at last received from England for the expansion of the Indian Army and for the increase of supplies and munitions of all kinds. So India's war effort began to go ahead in earnest. Congress propaganda failed to deter most Indians from taking advantage of the new opportunities offered. There was a rush of peasants to join the forces as sepoys, and of educated classes seeking posts as officers. Many business men turned speedily from financing Congress to making money out of war contracts.

Wavell's unexpected and brilliant victories over the Italians in North Africa, and the destruction of the Italian Empire in East Africa during late 1940 and early 1941, gave a big boost to morale. Two Indian Divisions had played a conspicuous part in these victories which aroused feelings of pride throughout India. The appointment of a very popular officer of the Indian Army, General Auchinleck, as Commander-in Chief, instead of the usual British Service officer, led to a change from some of the old ideas. For example that British officers could not serve under Indians. Auchinleck's tact and drive generated fresh vigour that eventually raised the strength of the Indian armed forces from around 185,000 at the outbreak of war to over 2,000,000 by the end. That these forces were entirely voluntary, and there was no conscription in India, speaks volumes for the relationship between Indians and the British.

The rapid and spectacular successes of the Japanese following the crippling of the American fleet at Pearl Harbour, and the sinking of HMSs "Prince of Wales" and "Repulse", brought the hitherto distant war unpleasantly close to India. Singapore surrendered on 15th February 1942. Rangoon was evacuated on 7th March and the whole of the Bay of Bengal lay open to the Japanese. It was a time of extreme peril comparable to the summer of 1940 after the fall of France. There appeared to be nothing to stop the Japanese from landing where they chose on Indias' long coasts. The position was so grave that Churchill is said to have warned the King that besides Burma, Ceylon, Calcutta and Madras could fall to the Japanese. This was the position during the first quarter of 1942 whilst we were being trained to command Indian or Gurkha troops at Belgaum. Indeed just before our course terminated we had news of large scale attacks on Colombo and Trincomalee, and on our shipping in the seas off Ceylon. There was considerable loss to our naval and merchant vessels, and some of the RAF planes were destroyed on the ground at Ratmalana, near Colombo. However the Japs were unaware that the Colombo Race Course had been hastily converted to serve as a landing ground for fighters of the Fleet Air Arm, and that only a few days before the first raid reinforcements had flown in from the Middle East. Some had been rushed overland through Iraq and Iran, and others had flown off an aircraft carrier several hundred miles west of Ceylon. Their timely arrival took the Japanese by surprise. Many Japanese fighters were shot down, helped by the surviving planes of the RAF. We also heard that the CPRC (or the remaining elderly reservists still available for duty), had taken part in repelling the raiders with machine gun fire. At Belgaum we almost wondered whether we were in the right place. Perhaps if we had stayed in Ceylon we could have been of greater use to the war effort. Such thoughts did not last long as the survivors of our army in Burma began to arrive in Assam, worn out, and many dying, after their extraordinary ordeal marching, retreating and fighting their way through the length of Burma in the most appalling conditions.

On Sundays several of us from the Ceylon Planters contingent would set out on our bicycles to explore the countryside. We discovered several of the old Maharatta hilltop fortresses. To reach these it was necessary to leave our bikes and clamber up steep and rocky slopes. The old forts we visited were all in ruins. They were very well sited to command the adjoining country, and seemed to be quite unvisited by the people in the villages below. Doubtless could the walls speak they would have told many an exciting tale of battles of long ago. On one occasion,when we had a free week-end, several of us set off into the jungle towards the coast. We came upon an isolated but well main-tained bungalow on a hill-top in the jungle. We persuaded the "chowkidar" (watcher or caretaker) to let us stay the night. This man seemed most upset and very nervous at our intrusion, but it was now dark and there was nowhere else to go. It was only in the morning that he managed to make us understand that we had crossed into the Portuguese territory of Goa and the bungalow was kept for the recreation of senior Portuguese officials. Presumably this meant mainly for shikar? Indeed we saw grey jungle fowl and spur fowl close to the building. Portugal was neutral and in the capital of Goa, Panjim, not far away were many interned German, Italian and some British nationals. Interned for the duration of the war. Not wishing to join them we beat a hasty retreat and did not halt until we were quite certain we were well clear of Portuguese territory. We found a river flowing through the jungle and, to cool off, we stripped and dived into a deep rock pool for a refreshing swim. Afterwards while we were drying off on the rocks a large crocodile put in an appearance. It was not our most successful expedition and we kept very mum about it when we got back to the OTS. As time went by and we were not hauled up before Colonel Pym, the Commandant, we presumed that the chowkidar had been satisfied with the bacsheesh we gave him and had not told his Portuguese masters about our unexpected visit.

In one of our Ceylon contingent platoons there was an exception to the rule that we were all from the CPRC. This was a very large and tough Pole who somehow attached himself to us. I forget his name but it was something like Wijinsky. Anyway the liking was mutual and Wijinsky became a sort of great cheerful bear-like mascot. Age wise he seemed positively ancient to us and he sported a crop of short wiry grey hair, though this may have been due, in part, to his trials and tribulations in getting away from the Russians and the Germans in Poland. His English accent was peculiar. He could not, at first, ride a bicycle. The file on each side of him would support him as we rode off to parades. We pushed him up hill and tried to restrain his progress during descents. Sometimes there would be pile-ups, when Wijinsky would emerge from a tangle of cycles roaring with laughter, much to the fury of the Company Sergeant Major, and to the detriment of good order and military disci-pline; though I suspect the CSM secretly enjoyed the entertainment as much as we did. I never learned how Wijinsky managed to get himself away from Poland, cross the rest of Europe into Turkey, then make his way across Turkey and Iran to India, or how he persuaded the authorities to accept him for Belgaum OTS. Nor did I ever learn to which unit of the Indian Army he was eventually posted. I suspect it was the Intelligence Corps. Certainly Wijinsky was not inexperienced in military matters. Wijinsky must remain as one of the thousands of mysteries of the war.

One day in the gym where I was enjoying practice on the horizontal bar, a P.T. instructor came up to me and asked if I would give him some boxing practice. I had not been in a boxing ring since leav-ing school, but I was at least a head taller than the PTI, so.felt it would look pretty bad if I refused. The PTI was a tough cocky little man with a barrel. like chest and muscular arms. He was soon battering me round the ring whilst, at the same time, exhorting me to "have a go" at him. This I was only too keen to do. In fact I wanted to give him a really good bashing. Eventually, in spite of his weaving and ducking I got in a couple of pretty good lefts to his face. Meantime he gave me some thundering blows to my ribs, always seeming to strike exactly the same spot. He was obviously a good boxer. Later I was told he was the army champion middleweight, and that he was well known for getting unsuspecting cadets into the ring to give them a hiding. So I was pleased that I had got a bit of my own back even though I suffered a lot of pain for many weeks. He had probably cracked one of my ribs. Later on when I was much fitter I tried to get a return match but never managed to tie the man down. Nowadays one hears quite a bit about bullying in the army and at schools. By far the best cure for bullies is to stand up to them and play them at their own game. Almost invariably this stops them for good.

At the end of our course when we had just been told of our individual postings, I played right back for Ceylon against an Indian Hockey side which included a couple of Sikh internationals. They had played in Indias' Gold Medal winning Olympic team at the 1936 Olympics. Just before the final whistle one of these fellows hit me a treme ndous crack across my left ankle when he was trying to shoot for goal. After a night of pain I reported to the M.0. He scarcely looked at my ankle and to my absolute fury insinuated I had nothing wrong; that I was trying to get hospitalised to avoid the posting to my unit. The idea had not entered my head. All I wanted was treatment so that I could get to the fleshpots of Bombay, with my cronies, for the week's leave before joining the newly raised 5th Battalion, 9th Gurkha Rifles, at the 9th Gurkha Regimental Centre, at Dehra Dun. So I told him to go to hell and bought some strapping with which to tie the ankle up. Fortunately we had no more boot wearing parades before I caught the train for Bombay with most of the rest of the contingent. It was a couple of months before I could walk without the bandage, and that ankle has caused me trouble ever since. It is an odd shape and I am convinced that there was a fracture.

Mike Horsefall and I had booked into what turned out to be a seedy hotel in Horniman's Circle as the Taj was chock full. After a couple of rotten nights Dick (R.C.P.) Adams very kindly arranged for two extra beds to be put into his room at the Taj, where we joined him. We thoroughly enjoyed the cool and comfort at this well run establishment. It was then the most luxurious hotel in India and, in spite of the war and the overcrowding, was early in 1942 still running very much on a peace time basis. Many of our contingent gathered in the bar and amongst those staying at the Taj were Sandy Richardson and Percy Grey. They, had commandeered a Sikh taxi driver and his taxi and paid him to keep it outside the hotel entirely at their beck and call for the whole week of their leave. No mean achievemnet in those war time conditions, especially as thay tended to use it at any time of the day or night. Since we had just become second lieutenants (emergency commissioned) in the Indian Army we were automatically extended the privilege of Honorary membership of many of the City's clubs. This included the Royal Bombay Yacht.Club situated just across the road from The Taj Mahal Hotel. There I was fortunate to be able to get in a bit of sailing in one of the club's Tom Tit handy 18 foot gunter rigged half deckers. These were easy to sai.l single handed so I enjoyed sailing out and round many of the ships anchored in the roadstead.

The short holiday in Bombay was not exactly a period of relaxation but it was a tonic and a big change from the strictly organised routine at Belgaum. Our time was our own; we had no idea what the future held in store for us. It was a case of "eat, drink and be merry......" We did not contemplate the rest. The hotel's bars and restaurants were always crowded chock full with officers of all ranks from all the services. In the evenings civilians in dinner jackets; Indian gentlemen smartly dressed in long coats with high collars, and jhodpurs; ladies in evening gowns or colourful saris; and always the neat looking servants, heads adorned with pagris, long white coats and pantaloons, and

*The author as a Second Lieutenant
with 5th BN. The 9th Gurkha Rifles, May 1942.*

brightly coloured cummerbands. It was difficult to believe that the Japs were hammering at the eastern end of India and that Dakka and Calcutta were being raided from the air. We visited Breach Candy and Juhu, driving past Chowpatty, the Towers of Silence and the Hanging Gardens of Colaba. We drove down Grant Road in a ghari Some may even have stopped to sample its delights.

Our leave came to its end. Three of us had orders to report at Dehra Dun, a military station situated in the Siwalik range; foothills of the Himalayas. Percy Gray was to report to the 3rd. Gurkha Rifles at Ghangora. Mike Horsfall and I were due to go to the Regimental Centre of the 9th Gurkha Rifles at Birpur. We caught an express of the Bombay Baroda and Central Indian Railway (BB&CIR), at Bombay Central Station and settled in for the 56 hour journey. The train headed north to Baroda. I was only to learn years later that my future wife's great grandfather was the General commanding the garrison and District in the 1870's, and her grandfather and grandmother the General's daughter met and married there at that time. After Baroda we crossed the Sindhiya hills and steamed north east through many places with romantic sounding names such as; Godhra, Donad, Ratlam, Mehidpur (where my future wife's grandfather arrived in a bullock cart in 1870). Then on to Kotah, Bharatpur and Muttra of pig sticking fame. Delhi was reached late on a really very hot and sticky afternoon. We had an hour there so took to the platform to escape the stifling heat of our compartment. A friend of Percy's, stationed in Delhi, arrived with a flagon of beer, but before we could do it justice a sand storm roared up and forced us back into the compartment and to close the windows and shutters against the flying sand. Next morning we awoke to cooler weather, in a vast well watered plain where camels were acting as draught animals, and buffaloes wallowed in muddy pools. we were in the granary of India.. The country changed again, becoming hilly and wooded. We were approaching the Siwaliks and the long climb through the Doon Jungles. The District of Dehra Dun lies at an elevation of about 2,000 feet, in the northern corner of the United Provinces, and forms a valley between the Jumna and the Ganges rivers where they emerge from the mountains. The valley is shut in on the east by the main Himalayan range, and on the west by the Siwalik range of much lower elevation.

I have no memory of our actual arrival, or how we got from the station to the regimental lines at Birpur, or anything about our reception, but I do remember,very well,approaching the Officers Mess for breakfast next morning. It was in the month of May and the path leading to the hall passed between two hedges of the most beautifui and sweet scented sweet peas.

I was allocated a room in No.14 Bungalow. The outlook was magnificent. It faced north across a valley. At the foot of the valley was a rushing stream and a narrow road running beside it. Beyond the stream ranges of jungle covered hills rose,one behind the other, forming a high background and a beautiful jagged skyline. High up in the distance could be seen some of the white toy-like buildings of Mussoorie hill station. The mess stood a few hundred yards away to the right. There were two other occupants of the bungalow. One was a pleasant, older officer who, I believe, was a missionary in peace time. The other was a fellow who looked young enough to be a schoolboy. John Bradbourne had been posted straight to the second battalion in Malaya, where he arrived with a reinforcement draft in the middle of the battle for Ipoh. From the moment of arrival he was continually in action until the battle at Slim River when he and a party of Gurkhas under Captain Hart became separated from the battalion and found themselves behind the Japanese positions. They decided to split up and make their way individually, hoping to rejoin the battalion further south. Hart and Bradbourne realised that their tall and obviously western appearance might hinder the men who were less easily recognisable. After many adventures John reached Sumatra and, eventually, arrived back at the regimental centre. He was one of only a handful of the battalion to get away. He was very upset by his ordeal and may well have felt a sense of guilt at not sharing the ordeals of imprisonment which became the fate of his fellow officers and men. John had a gramophone and a collection of classical records. My memory is of the strains of Bach emanating from the next room at almost all hours of the day and night. I owe my liking for Bach's music to John. After the war John became a missionary in Rhodesia. Towards the end of the troubles leading to the founding of Zimbabwe the mission was attacked by terrorists and John was brutally murdered. He had calmly tried to negotiate with the thugs and was cut down whilst trying to save the residents of the mission.

The regimental centre was under great strain at the time of our arrival. It was expanding at an unprecedented pace under very difficult conditions. Lt. Col. H.A.Fagan, the commandant, had, not only to organise, train, feed and supply a huge intake of recruits, but had also to build new accommodation to house them. As the recruits poured in - every one of them a volunteer - the problem of training and equipment was acute. At one time four men had to share one rifle. The men came straight from isolated villages in the Himalayas, often two weeks march from the recruiting depot. Few if any had ever seen a bus or a train. They were illiterate, and had to be taught to read and write in roman urdu script before they were taught military matters. The fact that they were incredibly keen to learn and all showed, amazing tenacity went a long way to overcome most problems. As a newly arrived British Officer I could not but be impressed with the energy, smartness and keenness of these wonderful men.

I found several good fellows amongst recently arrived officers. There were 'old china hands' such as Feathers Fetherstonehaugh and Bob Bailey, and Bill Thom, from Northern Rhodesia. There were even a couple of Darjeeling tea planters* In fact I was surprised to see tea growing by the roadside not far away. It was admittedly rather poor stuff but nevertheless it was tea.

I briefly met Lt. Col. C.W.F. Scott who was about to leave for South Africa having been invalided out of the army. Before the first world war Scott had been in the Ceylon Planters Rifles, and when virtually the whole unit volunteered for service in 1914, he went to the Middle East with it. The authorities were so impressed by the men and their bearing that every one was commissioned immediately. Scott took part in the initial landings at Gallipoli, where he was severely wounded. Later with B Company, 2nd. 9th G.R., he was again wounded. Scott had been planting on Kirklees Estate in the Uduppussellawa district, and played rugger for the Uva XV. To get to a match in the provincial capital, Badulla, he would walk or ride across some 25 miles of incredibly rugged hill country. After the match he would stay the night at the Uva Club before making his way back to his estate in the same way. It was a privilege to meet a Ceylon Planter whose name had been famous on the rugger field, and who had made his home with the 9th Gurkha Rifles. Other officers, and their wives, who were kind to me were Major T.A.Massie, Major E.H.Russell and Major J B.Peacock

I cannot recollect being allocated any specific task or duty during my first few weeks at the centre. I accompanied various Gurkha working parties, mostly in an observer capacity, to learn about the men and their language, and how things were done in the regiment. I was taken to the Gurkha Officers Mess to meet those marvellous men the Gurkha Officers. The rum always flowed like water. Many of the GOs were reservists who had rushed back to the colours from their homes in the hills. There was a tented hospital nearby. It had been established very recently to cope with the many many casualties amongst the soldiers who had marched and fought their way out of Burma. They had reached India in the last stages of exhaustion, worn out by disease and lack of nourishment. The strength of many had been so sapped that they simply laid down and died. I visited an old Ceylon

* Gorst and Shipton.

At Birpur estate, Dehra Dun, 1942.
9th Gurkha Rifles Regimental Centre

The Officers Mess

The Pipes and Drums.

No. 14 Bungalow, Birpur Estate

View from Officers Mess to Himalayas.

friend there. He was very ill but, I am glad to say, eventually recovered and after the war resumed planting in Ceylon.

In the evenings, several times a week, we would send for a tonga to drive us in to the Dehra Dun Club. When we clambered in the combined weight of three of us nearly lifted the miserable pony off the ground. The driver had to sit out on the shafts to counterbalance the weight. It has been said that a tonga abolishes all distinctions of class, race and colour. Nothing more undignified could be invented, but it was a cheap and fairly safe mode of travel, fast enough for our needs and, had a motor been available we would most likely have crashed it on the way back. I managed to join several shikar parties in the hilly forests close by. It was great fun to hear the clever way the Gurkha beaters directed the game towards the waiting guns, with a tap tap on a tree here, imitation of an animal cry there, and finally, when close to the guns, much hallooing and message calling. Later, I had several memorable week-ends at the forest bungalow of Rumpur Mandy, close to the rushing waters of the upper Jumna, where the melted snow waters emerged from the Himalayas,. On at least two occasions Bob Bailey and I borrowed a truck, loaded it with rations and our Gurkha orderlies, and spent the night under the trees outside the bungalow. Early in the morning we would leap aboard a timber raft which carried us down the rapids to a site reccommended by the shikari. We would then make our way back towards our, base shooting as we went. Our bag was mostly Jungle murgi, but we also shot the odd partridge and, on one occasion twenty-two peafowl were sent over by the beaters. Roast peafowl in camp, if well basted, makes a most tasty dish. Another delectable and sustaining.food is buffaloe curd. It tasted best when one met the herdsmen with their animals deep in the jungle during a long treck on a very hot day. The fresh curds always seemed to be icy cold in the earthenware pots. Another memory is rising very early- about 2 a.m.- whilst staying in a local zamindars' house, and riding through jungle and small villages in the back of a bullock cart. As we approached each village we would hear the musical call of the village watchman who were trying to prevent the fruit bats destroying the mango crop. Just before dawn we reached a beautiful park like vale. We stopped for an al fresco breakfast before starting the day's shoot.. The shafts of the carts were propped on the usual forked poles, and the draught bulls hobbled and turned loose. Bob Bailey, our Indian host and I took up strategic positions,,and the villagers commenced beating through the scrub and light jungle. I was loaded with number four or number six shot, for jungle fowl, when I suddenly saw a tiger standing looking at me about fifty yards to my right front. Bob was about 100 yards from me, so the tiger was probably half way between us. I kept very still and the tiger looked away and slowly crossed the open space and disappeared to my rear. His movements were smooth and graceful. Then a couple of peafowl flew over. I let them go not wishing to disturb that beautiful beast or to chance scaring him back towards us. He had appeared up wind of us. Bob had also seen him and kept absolutely mum. Whenever we went to Rampur Mundy we would see villagers crossing the rapidly flowing river on primitive rafts. These sometimes consisted of several inverted chatties (earthenware pots) lashed to poles. The air inside the chatties provided the buoyancy. Another method was for the villager to lie across an inflated buffaloe hide and progress by paddling with his feet and hands. The river may have been 75 to 100 yards wide but the.raft, if it could be called that, would travel at least a quarter of a mile downstream before reaching the further shore. on one occasion I decided to swim across. The water was icy and the current dragged me a long way, so I walked a good half mile up stream before attempting the return journey.

The enjoyable times described above were just short interludes in the midst of serious training, and the beginings of shaking the new batallion into a fighting unit.. Following the loss of the Second Battalion, along with the 2nd. Battalions of the lst., and 2nd., Gurkha Rifles, the C-in-C India, had requested from the Supreme Commander-in-Chief, Nepal, permission to raise three additional war substantive battalions as replacements. Assent was immediately forthcoming and, on 15th July 1942, the Fifth Batallion, 9th Gurkha Rifles, officially came into existence. Lt.Col.O.K. Steveney was appointed C.O. and Capt. J.B.Peacock his second-in-command. 21 British Officers, 18 Gurkha Officers and some 650 other ranks made up the strength. There was a sprinkling of experienced officers from other batallions, but the majority of officers and men were new arrivals at the regimental centre. We were told that our role would be as "revenge battalions", raised for the purpose of avenging the lost units in Malaya, but first we would have to be really well trained in order to carry out our alloted task successfully. This was a sign for much organised activity at getting sorted into four rifle companies and a Headquarter Company, each with its sub-units and various organisations. The officers, N.C.Os and men had to be assigned their duties, Battalion Standing Orders were drawn up; boots, uniform, equipment and weapons, distributed. Weapons were still short, as was motor transport. Housing and messing had to be provided and sports fields and equipment found, as well as all the multifarious detail needed to ensure a happy and efficient unit. There being no military transport available a fleet of clapped out old buses was commandeered. Our recruits,newly arrived from the distant hills, were taught the rudiments of driving and mechanical maintenance on these vehicles. Most of the local roads were narrow and twisty with steep khuds at the outer edge, so there were a number of unfortunate accidents.On one occasion when a bus broke down the Gurkha driver summoned another to tow him home. Assuming that the disabled vehicle would automatically follow in the wake of the tug, both drivers got into the front seat of the front bus,leaving the steering wheel of the other unmanned but some ten or twelve Gurkha passengers in the back. Inevitably, on reaching a bend,tne towed bus failed to follow in the tracks of the other, and crashed down the khud severely injuring some of the passengers. At least that lesson was never forgotten. During this period I was sent on a course to Saugor, right in the centre of India, in the Jungles of Central Province. I had to learn all about 3 inch mortars with a view to taking command of the mortar platoon of Headquarter Company. While at Saugor I went down with malaria and spent about ten days in hospital there. This was before catching malaria became a crime, and before the army had really got to grips with antimalarial treatment. The incidence of malaria became so bad that tremendous efforts were put into its prevention and control; to such an extent that a few months later infection became a crime subject to severe punishment.

Formation of the battalion was still incomplete when the internal security situation intervened. The politicians of Congress and some of the other parties demanding independence had seen the British defeated in Malaya and Burma, and the Japanese fleet in virtually total control of the Bay of Bengal. With such ineffectual resistance they had no confidence in Britains' ability to defend India. A considerable fifth column had come into existence,especially in Bengal where certain sections were suspected of being in contact with the Japanese. The general impression was, that in the event of a successful landing by the Japanese local Congress leaders would be the first to garland them and offer cooperation. Also, under Gandhi's influence, Congress now became more uncompromisingly hostile than ever before, in spite of an excellent offer by Sir Stafford Cripps of full independence on cessation of hostilities with an allied victory. Gandhi was convinced the axis powers would win. In August the All India Congress passed a "Quit India" resolution by a huge majority. This was immediately followed by an impassioned appeal by Gandhi who declared he wanted "freedom immediately, this very night, before dawn..... Do or Die. We shall either free India or die in the attempt; we shall not live to see the perpetuation of our slavery".

In the early hours of the following morning all members of the Working Committee, and most of the national and provincial Congress leaders were arrested, and the Congress committees were all declared unlawful associations. This swift action was taken with the unanimous approval of all the eleven Indian members of the Viceroy's Executive Council.

*Shooting rapids on a timber raft. See camping gear on board.
Bob Bailey watching our progress*

*Beaters near Dhaula Tuppar: Our orderlies Bhimbahadur and Perembahadur
at attention and carrying Peafowl, 1943.*

*Primitive method of river crossing, by inflated Buffaloe hide,
at Rumpur Mandy on the Upper Jumna, 1943.*

Fish netted by villagers on the Arsan River, near its junction with the Upper Jumna.

At such a critical moment of the war it was essential that the Government take every necessary measure to suppress mass civil disobedience and open rebellion with all the means available. The swift action took the trouble makers by surprise. In spite of this there was a widespread outburst of violence and sabotage aimed mainly against communications. organised gangs removed railway lines and sleepers, smashed signals and signal boxes, cut telegraph and telephone wires and uprooted their posts. Mobs attacked and burned railway stations and damaged bridges. Police stations were besieged, often by crowds thousands strong. Post offices and other government buildings were attacked. The attacks began simultaneously in widely separated localities. It was obvious that this was not just a spontaneous reaction to the arrest of Gandhi and the other Congress leaders. There was evidence that sabotage had been concerted and prepared in advance.

For some days whole districts were wrested from government control, and the railway was so severely dislocated that Bengal and Assam, and the armies protecting India's eastern frontier against the Japanese, were completely cut off.

The trouble started in August while I was away in Saugor. I had a somewhat interrupted journey back to Dehra Dun where I found the Batallion under orders for railway protection duties. The line between Hardwar and Dehra Dun was already being patrolled by men from one of our rifle companies and, I was told that, only a couple of days before a mob had set upon two BORs (British Other Ranks), in the station and had burned them to death. On my way to Saugor I had had a minor excitement but did not connect it with any political trouble. I was the single occupant of a two-berth sleeping compartment and on leaving Hardwar station I lowered the window to get some cool night air. As I poked my head out of the window I was attacked by a dirty, wild looking ragamuffin, who must have been standing on the carriage step. Probably he had hoped to remain there until I put out the light when he would have forced his way in and robbed or attacked me. I tried to raise the window with one hand whilst holding him off with the other, but he forced his full weight down on it and tried to clamber in. Fortunately I managed to grab my silver-headed swagger stick and belaboured him about the head until he fell or jumped off and rolled down the bank. The train may have been travelling at about 20 m.p.h. I raised the window and shutters, and closed all the locks before retiring to my bunk. I still have the swagger stick with not only the regimental crest, but several dents on it from the thrashing I gave that intruder.

I reported to the battalion office to be told that H.Q Coy. under Major Maguire - including the mortar platoon which still had no mortars,- was moving immediately to Saharanpur, an important junction where lines from the north and south join the main route between the north-west and Bengal and Assam, and so a vital point in the country's communication system. H.Q Company travelled to Saharanpur in the clapped out old buses referred to earlier. We set up the orderly room in one of the station waiting rooms. Another was converted into a cookhouse for the men while the officers used the station restaurant. An "armoured train," was made up for our use. This consisted of an engine with tender in the middle. To both front and rear of this a passenger coach was attached, then an open truck and, at each extreme, was a truck loaded with ballast. Thus a ballast loaded truck led the way whether the train moved forward or reversed.. The idea was that the ballast would take up the blast from any explosion on the line. The empty trucks were for the train-guard armed with rifles and light machine guns, and the passenger coaches were for stores, rations, and parties of men destined for protecting bridges, stations, signal boxes and such like. These parties had to be relieved on a regular basis. After a few days we decided we needed to be more comfortable and persuaded the station master to provide a sleeping car. This was drawn up in a siding adjoining platform No.1. When it had been connected to the telephone line and mains electricity Coy. H.Q. moved in to one end. The company commander, his two British officers and the senior Gurkha Officer, were each allocated a sleeping coupe.
Nearly all the trains which arrived at Saharanpur from the west were carrying large parties of troops

returning from home leave to their units at the eastern front. As the lines to the east had been sabotaged we had to order these men off the trains and set up makeshift accommodation for them. A camp was pitched on the maidan and the men were, as far as possible, separated into groups according to their unit, class or religion. It was quite a complicated business organising rationing and cooking for the different classes. Fortunately the soldiers were, almost without exception, extremely helpful and cooperative. I do remember one tall, moustached, very fine looking and highly decorated Pathan subedar from a famous frontier regiment, who made every possible difficulty and complaint. He was one of very few exceptions. We received much help in our difficult task from the Remount Depot. This unit whose normal task was to gather in horses and mules, and to prepare them for use by horsed units, animal transport units, mountain gun batteries and the like, did everything it possibly could to help us. Being based at Saharanpur, the Remount Depot had many facilities which we lacked. It sported several unusual equipages. One of these consisted of a smartly painted covered cart drawn by a team of four beautifully turned out mules. These vehicles were put at our disposal for transporting equipment, rations, water and the sick to and from the new camp. This camp grew daily larger from the steady influx of leave expired men. Fortunately for us several special troop trains, scheduled to convey whole regiments from the Punjab to Bengal and Assam, had been stopped before they reached Saharanpur.

One day a report came in from Deoband, a small town about 20 miles south, that the railway station was under attack. The "armoured train" was not available, being out on some other scare towards Roorkee in the east, so I was ordered to catch the next scheduled train going south, with a platoon under command of Jemadar Ganjbahadur Gharti. He was an experienced G.O. with over twenty years service. The men piled into the front coach while Ganjbahadur and I rode in the engine. When we arrived at the small Deoband station all seemed quiet and the station was nearly deserted. The stationmaster reported a mob had threatened to burn the place down, whereupon he had telegraphed his message and told the crowd that the army was on the way which caused them to decamp hurriedly.. He also reported that the telegraph wires had since been cut and he was no longer in communication with anybody.

I left Havildar Dharam Raj Khattri at the station with a few men, and set out with Ganjbahadur and the rest of the platoon to march the mile or so to the town. We spread out with several yards between each man,to make our small force seem as large as possible. The town appeared to be virtually deserted so we marched through the main streets to "show the flag", knowing we were being closely observed from behind the shutters. While in the town I met and.chatted briefly with the inspector in charge of the police kot. He kindly invited me to dinner at his house in the central square that evening.

When we arrived back at the station during the heat of the afternoon, we were delighted to find that Dharam Raj had not only organised hot tea and food for us but had also, of his own initiative, sent a message via the guard of a northward bound train, telling Maguire of the latest situation. At about 5pm., having told Ganjbahadur to organise sentries around the station, and to send out a search party if I was not back by an hour after dark, I set off,armed only with a walking stick, for the police inspector's house. He was a Muslim in a predominantly Hindu area, and doubtless had sound motives in asking me to visit his home; perhaps to shew the people that the military were his good friends. We ate together on a first floor balcony overlooking the small central square, and were waited on by his wife and daughter in purdah. The square was deserted and every door and window seemed shuttered. There seemed to be a tenseness in the air. I left on the return journey refusing the inspector's offer of an escort, and walked through the streets trying to look as if I was out for a normal evening stroll. Jemadar Ganjbahadur greeted my return with obvious relief and we enjoyed a good strong dram of rum together. I hope that our small expedition helped, to some extent, towards keeping the peace in the Deoband area.

Next day a section came out under an N.C.O. with a message recalling us to Saharanpur. Over the next five or six weeks I went out on several night patrols,following intelligence reports that badmashes were collecting in various villages near the railway lines. We always dropped off the "armoured train" well out of sight and sound, and quietly approached and cordoned the villages; but always the birds had flown, or were too well disguised to be identified. Time was also spent visiting bridge and station guards who had a tedious and lonely job. Casualties from malaria were fairly numerous, and several men were knocked down by trains, resulting in the deaths of two young riflemen. Maguire was a handsome and dashing Irishman, very popular with everyone at the railway institute. I got to know some of those excellent railwaymen and their families at the institute dances and "housie housie" sessions. John Masters in his book "Bhowani Junction" perfectly describes the Eurasian railway people. The Colonel Savage of his book was a close double of Maguire. At that time Eurasians virtually ran the Indian railways, and did so remarkably well.

As the situation gradually resumed normality the delayed troop trains started to pass through Saharanpur. It was quite impressive to watch carriage after carriage pass through the station, every window crowded with cheering Indian soldiers, often they shouted their war cries. Next the men from the transit camp were gradually cleared and the camp was packed up. However the last of our riflemen did not return to Birpur, and their badly interrupted training, until the beginning of December. By that time I was in camp at Raipur training with my mortar platoon. We had received our full establishment of weapons at last. We managed to get in a lot of live firing practice. Our camp was in a small plain surrounded by the steep forest clad hills of Tehri Garwhal and there was a plague of locusts in the area. One day when we were engaged in live firing practice there was a large cloud of those unpleasant insects high in the air between the mortars and the target. We fired a salvo of bombs which rose up into the locust cloud on their way to the target, but the locusts were so thick that the bombs detonated amongst them, like anti aircraft shells destroying thousands of flying objects.

During December I took a couple of days leave to meet my brother Louis in Delhi. He had just arrived in India with the British 2nd. Division. I had not seen him since early in 1935. While there we went to the cinema and found ourselves sitting immediately behind the C-in C. India, General Sir Archibald (later Field Marshal- Earl) Wavell. During 1943 Wavell was appointed Viceroy and was succeeded as C-in-C by The Auk (General Claud Auchinleck, later Field Marshal Sir Claud etc).

Suddenly orders came to move to Loralai in Baluchistan. The Germans had invaded and occupied huge areas of Russia. They had captured Leningrad, were round Moscow, The whole of the Ukraine and the Crimea was in German hands and they had reached the Volga at Stalingrad, and encircled the city. The Russians were resisting heroically but in the event of Stalingrad falling there would be virtually nothing to prevent the German armies dashing eastwards to the North West boundary of India. To counter this threat defences were being hurriedly constructed in the mountains of the Khojak Pass, about 80 miles north of Quetta. We were told that we would be held at Loralai as a reserve for the forces manning the Khojak. The whole battalion entrained at Dehra Dun on 27th. December and reached the railhead at Harnai a couple of days later. It was my first experience of a troop train. We were well organised and comfortable. The men thoroughly enjoyed the journey,patticularly the periodical halts in isolated sidings where everyone had to detrain and exercise with P.T.

The Fort at Loralai.

A 3 inch mortar in action.

etc., and a shave and make plenty of hot tea in degchies which were filled with boiling water from the engine. At Harnai we were picked up by 30 cwt. trucks, which did several round trips to get the whole batallion over the 60 miles of rough mountain roads to Loralai. It became very cold and I wished I had kept the woollen socks gloves and balaclava given to me by the Colombo ladies. Many parts of the road were snow covered and the trucks were all fitted with chains. On one side mountains rose to over 11,000 ft, and on the other to 8,000ft.

The small military station of Loralai is situated at about 4,500 ft, close to the so-called Loralai river. It was dry most of the time. It was in the middle of a wide plain between mountains rising to about 6,500 ft to the north and south. The men were housed in a Beau Geste style fort on the east side of the small cantonment, with officers quarters and a small bazaar to the west. We took over several empty quarters and the officers mess was housed in one of these. I became mess secretary. The mess orderlies were keen, smart and very cheerful, but had to be trained from scratch. Being used only to brass drinking vessels and metal or earth eating utensils it took some time and practice for them to handle our glass with care. Initially the breakages were horrific. The bitterly cold weather took me by surprise. I had not experienced anything like it before. It was much colder than an English winter, added to which I had spent seven consecutive years without leaving the tropics, so my blood must have been getting thin. Fortunately we were well supplied with,rum which to warm our cockles of an evening. This powerful brew was an Indian product and we ordered it by the barrel.

The months spent on railway protection duties had stopped our training before it really got started. So now we had to start again almost from scratch. Thanks to the keenness of all ranks an inspection by General Money, the Baluchistan District commander, gave the batallion a very favourable report that spring. For recreation we played six-a-side hockey on small 'matti' pitches, which were very fast and constructed with a small ledge raised round the ground so that the ball bounced back into play instead of going off the pitch.The football team beat all-comers to win the District Cup. One of our team was a ginger haired Yorkshire man. He became very popular with the locals, as well as with our men. The Pathans followed Reggie Gunter everywhere and would have done anything for him. His bright ginger hair made it almost certain, in their eyes, that he must be a descendent of The Prophet. Poor Reggie. He was fine in the cold weather, but the fine skin that went with his red hair meant that he suffered terribly from prickly heat in the hot weather. This became so bad that he had to be transferred to cooler climes. Volley Ball was very popular with the men and they became expert at it.Matches between platoons and companies engendered a great deal of rivalry and excitement.

An Animal Transport Company was stationed at Loralai and,thanks to its C.O., we had the use of some of his mounts. This officer was a great character and was most unusual, or rather unique in the Indian Army. Believe it or not, he was a Texan with a commission in the R.A.F. He had joined at the beginning of the war and became a fighter pilot during The Battle of Britain. He was badly shot up on more than one occasion. Finally he crashed and was severely injured, with the result that he was declared medically unfit for flying. When sufficiently recovered he looked around for further military employment and, being an expert horseman, had eventually been accepted for duties with animal transport. Such units being then only existent in India he had been sent out and was posted to Loralai. He set up a riding school for us 9th G.R. officers. Whilst in the riding school under his tuition I managed to come a cropper by making my mount cross his fore-legs whilst we were practicing a figure-of-eight. I had managed to get in some riding at Dehra Dun, but our Texan had much better schooled horses. He also organised cross-country paper chases on Sunday mornings which were great fun. Unfortunately, for us, the Animal Transport Company moved on, probably to the Arakan or Assam., After it left we organised 'Khud' races. The whole battalion took part

On Parade

Subedar Tikkaram Khattri, 2/, 'C' Company
and his brother, Subedar Tekbahadur Khattri, 2/, 'A' Company, Baluchistan 1943.

Presentation of medals at Loralai, Baluchistan.

'A' Company, leading the Batllion past General Money; the author in front, Loralai 1943.

Gurkha Officer 5th 9th G.R. Jemadar Kharakbahadur Mall.

127

in some of these. My best effort was to come in eleventh. That summer the gardens of Loralai were full of the most delicious fruit, much of it going to waste as about half the quarters were empty. I used to help myself to the most delicious apricots, peaches and grapes etc., that I have ever tasted.

It was at Loralai that I first came across the 'karez' system which is native to Persia and Baluchistan and provides permanent running water in arid country. A place is chosen near the head of a valley where the run-off from the hillsides collects far underground. Then a shaft is sunk just wide enough for a man to descend, until water in good quantities is reached. Then; without any instruments, using only the eyes; a tunnel is dug along a line which the water will naturally flow. This tunnel is entirely below the surface, but with a shaft to the surface every 150 yards or so, for the removal of earth, and for access for inspection. Thus a regular clean flow of pure water can be conveyed for many miles to a point where it can be brought to the surface for irrigation,for watering livestock or for any other purpose. I learned about the existence of karezes by falling down one of the shafts when out on the plain a few miles from Loralai. It was very cool down the shaft and the water tasted delicious and was icy cold. Later I came across many karez systems. One day, a very hot one, when we were on the way home after a battalion scheme we had fallen out and were resting by a track. There was a sudden, unexpected and very violent thunderstorm. As the rain died away leaving puddles of water on the bare ground, we noticed that there were thousands of tiny fish wriggling in the pools. It had literally been raining fish. The Political Agent stationed at Loralai would be invited by local chieftains, to shoots on their land. Sometimes one or two of our officers would accompany him. The shoots were quite strenuous affairs. We would walk up chikor and sisi partridge, golden plover, big fat pigeon, and the odd hare as we strode over the wide open hilly spaces. After some time we would come across a feast laid out in a sheltered hollow. There would be a white table cloth laid on the ground. On it would be spread platters of delicious looking pillau rice and mounds of fruit. To one side would be wood fires over which would be roasting joints of mutton. We would sit cross legged round the cloth and, after water had been poured to wash our hands, would tuck in hungrily. The mutton was deliciously tender. We would finish up with sherbert.

By the early summer of 1943 the Russians had broken out of the circle round Stalingrad and had started to advance into the Ukraine and the Crimea. This meant that the threat to India from the West had receded. Our training was altered and we went into camp for Frontier Warfare training, and on a brigade column to Dera Gazi Khan. The Baluchi tribes, although very independent and fierce, were not as wild and warlike as the tribes in the unadministered tribal lands beyond the North West Frontier Province, being given to smaller scale banditry and fighting. However there were prowlers after weapons round our lines and on one occasion an alert sentry slayed such a fellow. A few weeks later some bandits held up the dak lorry, robbing all the occupants and burning the vehicle. Captain Jim Fillingham, newly arrived to take over as adjutant, and Major Russel took out parties hoping for a scrimmage but the robbers slipped away to the sanctuary of the hills.

Towards the end of October 1943 the battalion was ordered to Fort Sandeman, the Heaquarters of the Zhob political agency, and the Zhob Militia. We were brigaded there with the 7/11 Sikhs and the 9/9 Jats under orders for intensive training. I had been promoted Acting Captain in August, and on first November became Temporary Captain. I was given command of 'A' Company, with Subedar Tekbahadur Khattri as my second in command. Jemadar Ganjbahadur Gharti was promoted to Subedar and became my senior platoon commander. The other platoon commanders were Jem. Kharakbahadur Mall and Jem. Dharam Raj Khattri. All were men of long and distinguished service with whom I was proud to be associated.

CHAPTER TEN

FORT SANDEMAN CHAMAN WANA.

"The man on the Frontier sees but his own square on the chessboard and can know but little of the whole game in which he is a pawn."

Colonel Algernon Durand.

The battalion's transfer to Fort San.deman in October 1943 involved us being brigaded with the 7/llth Sikhs and the 9/9th Jats. We took over our lines from a battalion of the Royal Gharwal Rifles which was posted to the Eastern front. The station also housed the headquarters of the Zhob.Militia whose men manned small forts or "posts" sited at strategic points in the district, and patrolled from them over the rugged country. The Scouts came directly under the control of the Political Agent in charge of the Zhob District whose headquarters and residence was at "Windsor Castle", a neo-gothic erection perched on top of a steep hill. The previous PA, H.A.Barnes, had been shot dead in his office in October 1940 by a disgruntled levy passed over for promotion. Barnes was the third Zhob PA to be assassinated in recent years.

Fort Sandeman named after Sir Robert Sandeman, was a small military station. in northern Baluchistan and lay in the Zhob river valley. All around were barren rugged mountains rising to 7,000 feet. The Afghanistan frontier was about 30 miles to the West, and 50 miles northward the Gomal river separated Baluchistan from the North West Frontier's tribal areas. Communication with Quetta was maintained by the twice weekly narrow gauge (2ft.6ins.) railway service, supplemented by a gravel road which ran alongside the line. Both were subject to interruption by storm washouts in summer, or snow drifts in winter. Other rough roads ran south to Loralai, and north through the mountains to the Scout post of Gul Kach on the Gomal river. The Zhob Scouts would drive along dried up river beds to reach some of their isolated posts such as Manikhawa, Ahmadi Durga, Shingar and Moghul Kot. I should mention that the army was only called upon for full scale operations in the event of the situation becoming too much for the Scouts to handle. The Scouts commanders were officers seconded from the Indian Army and such appointments were much coveted.

Although the country appeared rugged and barren there was abundant bird and animal life to be found by anyone interested. The best shooting was provided by chikor, sisi and golden plover, whilst during the migration season snipe and teal could be found for short periods. Chinkara, urial and Markhor roamed the mountains and foothills, but these tiny deer, and mountain goats with corkscrew horns were too delicate or beautiful to hunt down and usually remained safe in their craggy retreats. The country was ideal for military training

After a few weeks of training and practice at mountain warfare the battalion was sent out on a series of columns, piquetting the heights to allow transport etc to pass safely through. In one of these exercises, called ZHOBCOL, we spent several days piquetting the route north to Gul Kach on the Waziristan border. I liked nothing better than these strenuous columns where one had command of one's own Company, with a fair degree of autonomy, especially in sending platoons and sections to and from piquets, arranging covering support, routes, communications, and so on. At night we would bivouac in the open, behind hastily constructed stone perimeter walls, in the company of our ever cheerful Gurkha riflemen who relished the change from the routine training around the barracks and firing ranges. They invariably brought up the old joke about finding some soft stones for the saheb to sleep on.

The best part of the day, or night, was when the defences had been completed, and men and animals fed and watered, night piquets sent out to their positions on the surrounding hills, sick parade over and duty personnel posted. I would retire to my'own bivouac where my orderly would arrive with a mug of hot spiced tea, most likely laced with rum, and a large sweet parata. Then Subedar Tekbahadur, my company second-in-command, and the platoon commanders, Company Havildar Major, and Quartermaster Havildar would arrive out of the gloom to discuss the day's events and orders for the morrow. The back-drop was a starry sky, black mountains and the occasional flashing of a signal lamp dot-dashing a message to the camp. There would be sounds of muffled laughter from the men's bivouacs, and the stamping of hooves and jingle of harness from the mules tethered nearby. On one column we had a camel transport company of the Indian Army Service Corps with us, I was sleeping close to their lines and did NOT enjoy their roaring and grunting close above my head. It seemed to go on all night. I had visions of them breaking loose, or being cut loose by raiders and stampeding over me.

On the approaches to Gul Kach we passed through a group of nomadic Powindahs travelling south with their camels, sheep and yurts. When we drew level with one of the camels whose rider was clad in a burqa'a, a female voice suddenly called out "Say Digger, What about a fag?" There was no mistaking the Australian origin of the speaker. Bob Bailey quickly responded with a cigarette, the burqa'a was thrown back and a light passed. Before proceeding on her way the woman told us that she hailed from Melbourne and had been brought to Afghanistan by a camel trading Powindah about twelve years before.

Lt. Col. Steveney was ill at the time of ZHOBCOL and Major T.A. (Tam) Massie reported from lst. battalion to take command during the exercise. Massie was a most pleasant fellow and, to my mind, much more of a practical soldier than kind, scholarly Steveney who was a linguist with years in staff appointments.

It was a year and a half since the raising of our unit and we seemed to be no nearer to our declared goal as a "revenge" battalion for the loss of the regiment's second battalion in Malaya. Not surprisingly some of us were becoming impatient at being sent to wander round Baluchistan when we had joined up to fight the Japanese. Several of those new to the East found the long period of isolation in barren country, thinly populated by people they looked upon as savage aliens, devoid of "civilised" amenities, lack of female company, and harsh climate, hard to bear. On the other hand those of us who enjoyed the wild and beautiful country, liked contacts with the tribal people, the sporting opportunities, the company of our Gurkhas etc., were reasonably content, presuming that it would not be very long before we would be regarded as ready for battle in Burma or Malaya. In any case; all the signs were at that time that the war in the East would continue for several more years.

There was a small club within close walking distance from our lines, which was open to all the garison's officers. We often gathered there for drinks, especially during the long cold, dark winter. Liquor was strictly rationed apart from Indian rum which the battalion got by the barrel. There was also the Gurkha Officers Mess, where rum was very generously dispensed when we visited as guests. In fact the GOs enjoyed trying to get us BOs very drunk. They would lay on a stretcher party to take home any who succcumed to their wiles. Whatever the result it was a point of honour to be on parade, immaculately turned out, early in the morning. There is no doubt that many of us drank too much. Fortunately most of us were very fit and there was plenty of opportunity to sweat out the excess alcohol.. I always remembered that the Fort Sandeman Club had, only twenty three years before, been the scene of the slaughter of several officers during the tribal risings which followed the third Afghan War. The terrible events of those times were also plainly evident when we camped at Kapip, where an army convoy had been annihilated. The ground was still littered with horse and mule bones and the debris of battle.

The Brigade Commander, from a Sikh regiment, had a dislike of Gurkhas and took every opportunity to make our lives difficult. His dislike was compounded when our-men thrashed the Sikhs at football and the khud race. He even recommended our removal from Fort Sandeman to Chaman, and the replacement of our commanding officer.

At this time I was sent to Karachi for a "watercraft" course. The journey from Fort Sandemain to Quetta, via Bostan, on the 2ft6ins. guage line took about twenty hours. I was warned by friends to take a store of food and means of keeping warm, so I went equipped with an oil stove, eggs, bacon, bread,tea etc., and put plenty of blankets and sweaters in my bedding roll. The carriages had no heating or insulation but, otherwise, the single bunk compartments (The lack.of height in the carriages did not allow for a top bunk), were comfortable each being fitted with a small bathroom, washbasin and WC. Even so it was bitterly cold when we reached the summit at over 7,000ft. My train made good time and I arrived at the Chiltan hotel in Quetta in time for dinner and a comfortable night, prior to catching the Karachi mail next morning. I had much better fortune than our new second-in-command when he travelled the same route later that year. His train was stranded in the snow for two nights with the engines frozen up. He had to sit in darkness as someone had, as so often happened in those days, stolen the electric light bulb. He had omitted to pack the usual spare in his baggage. The track was eventually cleared and the engines re-started, giving him visions of reaching the comforts of a warm mess at his destination before dark. Unfortunately this proved optimistic as the leading engine ran off the line and, in spite of carrying jacks and other equipment to deal with such accidents, another night was spent in the cold dark compartment..

The others on the watercraft course were all men whose units were resting after returning from the first Chindit operation. The course was all about teaching make-shift ways of crossing rivers and swamps. It might have been better to have held this course before rather than after going into Burma, but perhaps, to give the organisers the benefit of the doubt, they were imparting lessons learnt during the operation. If so they never said so. As it happened this course was the only one when I got a 'D' (Distinguished) report. Probably because I was the only candidate able to swim a 500 yard wide tidal creek in full kit, boots and rifle. I was fit from Baluchistan whereas the others were probably unfit and recovering from the frightful hardships of Wingate's expedition.. All this seemed to be a sign that the 5th. 9th might soon be sent off to jungle warfare and that I had been selected to impart watercraft skills to our men. However this was not to be. Strange are the ways of the army. On return to my unit I learned that we were under orders to move to Chaman; just about as far as possible from any jungles with their swamps and rivers.

On the way back there was another amusing railway incident. Amusing for the onlookers but not for the victims. After leaving Karachi the train crossed the Indus at Hyderabad, traversed the barren land to reach Sukkur, where we again crossed the Indus and the desert to reach Jacobabad, one of the hottest places on earth in summer. During the following night we stopped at various small stations and ascended the impressive Bolan pass. I woke early in the morning to find the train halted at the little junction (whose name I forget), where a branch line takes the weekly train westwards to Zahidan on the Persian border. I got out to stretch my legs in the deliciously cool mountain air, noting that all was quiet in the other first class compartments. Not for long however. There was a loud howl from further up the train Then several agitated voices were heard complaining loudly, followed by a door opening and four agitated Europeans in various forms of night attire falling out onto the platform shouting for the guard and stationmaster. It transpired that they were four officers who had just arrived from the Middle East and were on their way to the Staff College at Quetta. They had taken the train in Karachi wearing uniform and had with them all the usual equipment and tin trunks etc. All - except the night attire they were wearing - had disappeared during the night. Probably they had failed to lock the doors or windows and thieves had quickly taken advantage and made off with every article in the compartment.. The theft had been skilfully carried out without disturbing any of the sleeping men. After much argument and recrimination the victims were persuaded to re-enter their cabin and the train resumed its journey for the 20 or so miles to Quetta. There the unfortunate new comers were seen leaving the platform in their night attire, baggageless, making for the rank of pony drawn tongas which always waited outside the stations in the way taxis wait here. I would dearly liked to have seen their reception at Staff College.

The onward journey to Fort Sandeman gave me time to reflect on the two years which had passed since I had come to India. This had more or less coincided with the fall of Malaya and Singapore. The retreat from Burma had quickly followed. During this time occasional snippets of news about my old planting friends had seeped through to me. Some of this came from Mike Horsfall who was still with the battalion, but he would be leaving very soon on transfer to the 5th. 1st. Gurkha Rifles at Razmak in tribal territory on the North West Frontier. That battalion had also been formed to avenge the 2nd.1st. Gurkha Rifles lost in Malaya.

Noel Dyson-Rooke and Ian Mackenzie, from Ury and Gonakelle estates, were serving in the same unit of the Jat Regiment in Malaya when the war with Japan started. The latest news was that Noel had got away, had turned up miraculously in Ceylon, and had joined the Royal Indian Navy and continued to fight the Japs at sea. It was only after the war, when I returned to Ceylon, that I learned some of the details. When news of the Japanese attack reached the Jats they, in accordance with the prearranged plan, moved across the border into Siam. They had some initial success but soon found that there were strong enemy forces behind them, and that more were landing further down the coast, cutting their lines of communication. They turned to fight their way out and during the course of the fighting Noel and some of his men were ambushed. Noel realised it was vitally important to get news of the Jap ambush location to Btn. H.Q. He got hold of a motor bike and drove straight through the ambush under a hail of fire. When he got to H.Q. he saw Ian Mackenzie there. Ian once told me that Noel's motor cycle was so shot up that it was amazing that he got through and that Noel was not even scratched. Later on, further south, Noel was again separated with a few of his men. They got hold of a boat and travelled south down the coast. When they thought they were clear they landed in hopes of rejoining the unit, but ran into more Japanese fire, so they hurriedly re-embarked and eventually crossed the Malacca Straits to Sumatra. After various adventures in that Island he joined others by taking a local boat and sailing her to Ceylon. Ian with the remainder of the battalion continued the retreat, fighting off the enemy as they went. They reached Singapore island and, on orders from the commander of the forces, surrendered with all the other troops. Ian was interned at Changi Jail under cruel and harsh conditions until later in the war he was sent to labour on the infamous railway under even more ghastly conditions. I learned all this from Ian himself much later. Both returned to planting after the war, married, and had families. Noel died in tragic circumstances whilst manager of Great Western Estate. Ian returned to England to Felixstowe, in about 1971, and died in 1974. His years as a POW had affected his health. I still dont know what happened to Jim Vokes and Neville Rolfe who were also in Malaya.

I had news that Walter Ogilvy, my old friend of Yapame days, was back in India having walked out of Burma. His story, which I learnt a bit about much later, had a rather better ending than Noel's and Ian's. Walter left Ceylon before the war to take charge of his uncle's rubber property in that narrow strip of Burma which runs down between the Bay of Bengal and Siam, South of Moulmein. In about March 1942 when Japanese army units were landing on the coast he was instructed to report to Rangoon immediately. He explained the situation as best as he could to his servants, telling them that he would definitely be returning, in due course, with British soldiers to chase the enemy away. He asked them to take care of his belongings and the estate during his abscence. "Don't worry, Thakin" they said. "We will look after everything". They buried Walter's silver in the garden, and took his fridge and other belongings, including his dog and golf clubs, into their own houses. Then Walter who was always a keen walker set off on what was to be the longest walk of his whole life. He found Rangoon city being evacuated by our forces, so he took a boat which was already over-loaded with refugees, and travelled up the Irrawady, and then up the Chindwin to Kalewa, where he joined the long columns of fleeing civilians finding their way through the wild, jungle covered Hill Tracts into Assam. Thousands died from disease and starvation on the way. After the war, following the re-conquest of Burma by Indo-British forces, Walter returned to his estate. He was joyfully greeted by the staff and workers, who produced his silver, fridge, golf clubs and other personal belongings. All were in good order, He soon had the estate working efficiently, but the political situation became more and more unstable. Then his wife died. So he decided to leave Burma and return to Ceylon, the land where he was born. Later on he married again, had a family and took charge of Attempitiya estate in the Uva Province. He died a few years ago.

Chaman was another isolated military station in Baluchistan. The Frontier and Customs post for Afghanistan was situated at the northern edge of the cantonment, which comprised a hospital, Messes and Officers quarters. The troops were housed in several small forts. There was also a tiny bazaar which catered for camel caravans, and a few dilapitated lorries bringing produce, mostly fruit from Afghanistan. There was a railway from Quetta. This terminated in Chaman and had been built for strategic purposes after the 2nd. Afghan War (1878 to'80). This railway, and a rough road over the 8,000ft. Khojak Pass, comprised the only communication with the rest of India. During the winter the Khojak Pass was often blocked by deep snowdrifts, whilst the railway through the Shelabagh tunnel was also impassible at times. The gradient from Chaman up to the summit (over 6,000 ft.) was very steep. When ice formed on the rails the trains would be unable to make the grade in spite of using three engines. Sometimes we watched, from a vantage point on the pass, to see if the train would emerge from the summit end of the tunnel., for occasionaly the crews would be overcome by

Top brass at Chaman on the Afghanistan border

Left to right: Two Staff Officers: Brig. Jenkins; Gen. Sir Claude Auckinleck, C in C India; Author: D. Shorter (Adjt); Major A. A. Mains (2½) and men making rafts.

Men of the Bikanir Ganga Rissala and their mounts.

the fumes whilst in the tunnel, and the whole train would run back down to Chaman with the unconcious men on the footplate. Usually there were two trains a week, except during the fruit season when extra trains were laid on. The Shelabagh tunnel was the longest in India, and the summit the highest on the Broad Guage. Being situated in the plain north of the Baluchistan mountains, Chaman was geographically in Registan, a vast desert, roadless and mainly uninhabited, which stretched away to Afghanistan's border with Iran. We were fortunate in having plenty of water. This was brought by aqueduct all the way from springs high in the mountains. The cantonment itself was quite an oasis thanks to the long established Indian Army custom that every unit stationed in a hot or barren area should plant shade, and flowering, trees and shrubs in their lines.

The other unit in Chaman was the Rajah of Bikanir's Camel Corps (the Bikanir Ganga Rissala), with whose Rajput soldiers and officers all our ranks established very friendly relations. Their camels were fine animals, specially bred for speed and endurance, and they claimed to have recently won a wager with a British Armoured Car unit in a race through the Scind Desert. They told us that a tumbler of water held in the hand of a rider at full gallop would not spill a drop. I don't remember if they proved this, but their camels gave a very smooth ride when at full stride..

Soon after we arrived at Chaman in the summer of 1944 we got a new CO. Lt. Col. H.H.W. (Harry) Rich came to us from the 4th. Gurkha Rifles. He was a bit of a fire eater. He might shout and curse at everyone, using the most colourful language on parade, but off parade and in the mess he could not have been more kind, friendly or charming. He instilled his keenness into all ranks and gave us all a good shake up. His method was to keep everyone intensely busy at all types of training, to devise realistic exercises, surprise alarms, and give all ranks a chance to use initiative, and to reach their own decisions. He tore strips off anyone who was slack or inefficient, but gave lavish praise whenever deserved. He knew how to handle men.

Shortly after Harry's arrival a new second-in-command was also appointed. This fellow had been through the retreat from Burma; then had worked in intelligence in Assam which he did not let us forget. He held a regular commission in the 9th, and may have been disappointed with his posting and may have been better as a Staff, rather than a Regimental officer. Overall we now had a better balanced set of officers than at anytime since the battalion's formation. Unfortunately Bob Bailey had left, but we still had the Rhodesian Bill Thom for Quartermaster, with Peter Harper understudying him. Nash had taken over as Adjutant from Jim Fillingham who now commanded 'B' Company. 'C' Company was under David Shorter, and another regular commissioned officer, Peter Richardson had 'D' Company. Grant and Edleston were both with HQ Company. Companies took it in turn to go into camp at Roghani, a small natural oasis some fifteen miles to the west of Chaman, in the foothills. Whilst in camp we would carry out section and platoon training in the mornings, then change into scruff order and enjoy shikar for the rest of the day. Communication with Bn. HQ in Chaman was by heliograph. However we were delighted whenever haze or dust storms made this impossible. Communication within the battalion was by flag, heliograph or runner. The heavy wireless sets then available were used between battalions and brigade. Only later that year were we to receive our first back pack wireless transmitters.

When carrying full packs, arms and equipment the march across the desert to Rhogani in hot weather could be quite exhausting. Mules carried the camp cookers, tents and other heavy gear. On one very hot day Jim Fillingham's brindle bull terrier decided to seek shade under a mule as it marched along. The dog was lucky to escape serious injury when the mule objected by lashing out and sending it flying. After strict water discipline on the march it was a delight to reach camp and gulp down the almost ice-cold water we had brought in chaguls slung from our packs and on the mules. Evaporation in the very hot dry air caused the water to cool in these slightly porous canvas containers.

Several small camel caravans passed close to the camp, travelling south from Afghanistan and beyond. These were very colourful and their people very primitive. Some of them carried exquisite Bokharan rugs and beautifully embroidered silks from Turkestan, which could be bought for a song. Thinking we would soon be sent into the war zone, and not wanting to be encumbered with luxuries, I resisted the temptation to buy. A decision I later much regreted as it was a chance in a lifetime. Local dacoits sometimes played a few tricks, such as stealing telegraph wires along the railway, moving border marking stones to their advantage, and damaging the pipe-line carrying Chaman's water supply. Once a mule broke loose from the lines in West Fort and bolted across the frontier into Afghanistan only a couple of hundred yards distant. Its muleteer, Rifleman Kharakbahadur, gave chase without hesitation. The mule returned of its own accord but Kharakbahadur was siezed and imprisoned. His release was obtained only after high level intervention by the Political Department in Quetta. When he returned Kharakbahadur was overjoyed to find his mule safe and sound. Nothing else mattered.

When the C-in-C India Command, General (later Field Marshal), Sir Claude Auchinleck paid us a visit I was able to demonstrate my company's skill at making boats out of charpoys and groundsheets and other improvised materials. However there was no water on which to practice handling them or testing their bouyancy, or lack of it. The general was much admired by all ranks of Gurkhas for the way he spoke to them, and for the fact that his service in the Indian Army was longer than any of our pensioners who had returned to the colours for war service.

I was sent to Kakul on a Frontier Warfare course. This involved another long train journey, more or less round three sides of a rectangle. The twice weekly train to Quetta left Chaman at 0600 hours and stayed the previous night in the station, so I decided to spend the night in my coupe after having dinner in our Mess. I was the only occupant of the sleeping car and woke early to find a large bearded, turbaned figure in the act of removing my suitcase from under my bunk, When I sat up he dropped the suitcase and fled out of the open door. I dashed after him but with bare feet could not run along the sharp ballast between the lines, and he got away. Perhaps I was lucky not to be stabbed by the pointed dagger he wore in his waist-band.

After staying a night in Quetta I caught the Lahore Mail, travelling down the Bolan Pass to Sukkur on the Indus, via the hot hell-hole of Jacobabad. From Sukkur the train ran North East to Multan, Montgomery, and Lahore, involving another night in the train and a night in Lahore before catching a train to Rawalpindi. At Lahore I spent the night in the seraglio of a nobleman's palace. The palace had been lent to the army for accommodation of officers for the duration and the only inhabitants were a few officers and servants. Yet another train conveyed me to Abbottabad (4,500 ft.) whence there was a mountain road to Kakul which is close to Murree (7,000 ft), the home of India's most famous brewery. The hill country was ideal for walking and in my spare time I found several vantage points giving magnificent views of mighty snow-capped peaks. There was Nanga Parbat (26,666ft) and the Karakhoram range. Views I will never forget. Abbottabad was the depot of the 5th. Royal Gurkha Rifles, a pleasant Hill Station where lived Europeans of all ages. It came as quite a surprise to see whole families about the place after my long monastic life in the outposts. Even more of a surprise was to find dancing going on at the club to the music of a Gurkha band.

During the autumn of the same year I went on another course. This time it was a Tactical Training Course at Dehra Dun, and was followed by a visit to the 29th. Gurkha Jungle Training Battalion, under canvas at Badshahibagh. The battalion was commanded by a well known character, Lt.Col. R.B.(Robbie) Fawcett, and was sited not many miles downstream from my old hunting ground at Rampur Mandy, on the upper Jumna. My position on this attachment was peculiar and unsatisfactory. I was not part of any course and had no official posting, and was given no responsibility. I merely tagged along as a sort of observer attached to BN.HQ. The Gurkha recruits under training at that time seemed very young and raw, compared with our "old soldiers" of at least two years service in the 5th. Bn. I went on a scheme which involved a company sleeping out in improvised bivouacs in the scrub jungle along the banks of the Jumna. There was a late severe burst of the monsoon, the river was swollen in flood, and rain poured down throughout. It was quite refreshing after the long dry years in Baluchistan. At the end of this rather abortive scheme I joined two young officers in riding the flood waters for the five or six miles back to base. We were carried rapidly down stream, sometimes clinging to logs and, in any shallows, pushing towards the far bank, before being swirled onwards again. That bank proved to be under water with currents pouring round the trees, so we made our way back to the higher east bank in similar fashion. It was an exhilarating swim.

The winter of 1944/45 was exceptionally severe. At Chaman we suffered many blizzards, and worse were the bitter blasts blowing straight off the Hindu Kush. After, several of our men fainted, on parades or at sentry duty,orders were received for sentries to be relieved half hourly. They were issued with posteens, gilgit boots, and sheepskin gloves. Charcoal braziers were kept burning near guard rooms and, owing to the extreme cold and hardness of the ground, sand bags were placed in suitable spots for sentries to stand on. Reports were received of benighted camel caravans being found with both men and beasts frozen to death.

At about this time liquor was in such short supply that the mess ran almost dry, which was a great calamity. Then news came of a shipment reaching Phipsons, the wine merchants, in Karachi. Telegrams were despatched, followed post haste, by a three ton lorry with an officer and four armed men as escort. They were joined by another three tonner from a sister Gurkha unit on a similar mission. The vehicles duly reached Karachi and collected their loads of whisky, gin and beer. On the return journey they were snowed up on a mountain pass. As it was bitterly cold our British officer told his men to sleep under his lorry, having made a snow wall round its body. The Gurkha officer in charge of the other vehicle told his men to sleep in it, on top of the bottles, and he joined them there. They all spent a cold and uncomfortable night, but next day it was found that the bottles which had been left uncovered and exposed in our lorry, had all frozen and burst. The bottles which had been warmed by the men's bodies were all intact. There was much recrimination when our vehicle returned with less than a quarter of its consignment undamaged. The laughter and rejoicing in the rival camp was great.

A message came from my brother Louis, with 2 Div. in the Deccan, that he was due to attend a course at the Staff College in Quetta, and he hoped I would get over to see him. I was determined not to miss the chance while he was so near and searched around for an excuse to go to Quetta on duty. An opportunity arose when a large stock of surplus blankets had to be sent off. The C.O. and the 2 i/c were both away but I got permission from the Officiating Commandant to move the stores in a small convoy. I used a tracked bren gun carrier to lead the way through the snow, A four-wheel drive carrier came next, to press down the tracks of the first. I followed in a three tonner, wheels chained and packed to the roof with blankets However the weather deteriorated and the leading carrier bogged down in a drift near the top of the Khojak Pass. The second vehicle got into difficulties trying to tow it out. I managed to turn my lorry round in a sheltered dell a few hundred feet lower down.

Order for leave to UK

FROM   BALDIST   25 10 20

TO   ZHOB BDE   **IMMEDIATE**

A.1202 (.)   CONFIDENTIAL (.)

MAY QUOTA STIFF (.)   Our 271/31/W/A of 6th APRIL (.)

MAJOR BROOKE SMITH to arrive TRANSIT CAMP KARACHI NOT repeat NOT

later 1st MAY (.)   Baggage 65 (sixty five) pounds (.)

Ensure officer vaccinated against smallpox (.)

I.A.O. 36/S/44 and G.H.Q. (I) 03 M/105/Org 10 Of 14th APRIL

to be strictly complied (.)   *not recd.*

Movement Order must show STIFF MAY QUOTA 1945 N.W.ARMY and

authority move ARMINDIA 81653/Q MOV 6 A 22nd APRIL

Confirm despatched

                        5/9 GR.

TO S/C (2)   *Forwarded for*   *W Haworth Capt.*

ER/25   *necessary action*   Staff Captain, Zhob Brigade.

⬭ **CONFIDENTIAL**

(i)   Number, rank, name and initials. *E.C. 7464 Major C.F. Brooke Smith (Indian Se*

(ii)   Regt. or Corps. *9th Gurkha Rifles*

(iii) Reason for proceeding to U.K. *'STIFF' MAY QUOTA 1945 N.W ARMY*

(iv)   Authority for proceeding. *ARMINDIA 81653/Q MOV 6A 22 APRIL*

(v)   Date of arrival overseas (present tour) *5 Nov 1935*

(vi)   Reporting instructions on disembarkation (if known).

(vii) Signature of individual for identification purposes. *A Brooke Smith*

viii) Signature, rank and appointment of despatching officer. *Mabel Hare*
                                                                Comd 5/9 G

(ix)   Official stamp of despatching Unit.

6TH BN, 9TH GURKHA RIFLES
26 Apr 45

Young Scott and three Gurkhas volunteered to stay and guard the carriers while I went to get help.Obviously nothing could be done in the dark, but at first light next day we got a company of Gurkhas with ropes and shovels and dug and pulled the stranded vehicles out. Scotty and one of the Gurkhas were frost bitten in spite of the extra blankets and the oil stove I left them. They melted snow over the stove,but it froze they said, as they poured it from their mess tins. It was, of course, against all the rules to leave men and vehicles out at night in that part of the world. Fortunately the exceptional cold kept any possible marauders away.We got the vehicles back to Chaman before the CO and 2i'c returned. Unfortunately the cylinder block of the second carrier was cracked and this had to be revealed, but I don't think they ever discovered the full details of this incident..It may be of interest to note that after I married, many years later, I learned that my wife's grandfather was in action at this very part of the Pass during the second Afghan War, whilst serving with the 9th. Bombay Native Infantry. It was some time before the pass was clear and I was able to reach Quetta and meet brother Louis.

Towards the end of the winter the battalion took part in what we called The Chagai Column. We were supported by an armoured car squadron of the Gwalior Lancers. A detatchment from the 2nd. Royal Lancers was also involved. We travelled as motorised infantry from Quetta to Nushki, on the road to Iran, to rescue a Political Agent. Beyond Nushki towards Chagai we moved on foot to make dashing infantry assaults and carry the commanding heights. According to the Regimental History, the District Commander declared he had never seen men move with such speed and assurance across the hills. My 'A' Company had the longest and most ardous cross-country patrol or 'gasht' to use the local term. The nights were still very cold, with hard frosts, but the days were sunny and warm. Several of the men of the 2nd. Lancers took pets with them in their armoured cars. One small dachshund was regularly used at the foot of its owners sleeping bag as a hot water bottle. Inside the bag. This country was barren by any standards, vast areas being uninhabited except by scattered bands of marauding smugglers who found plenty of room to hide amongst the rugged hills. The 2nd. Lancers fresh from their successful role, at long range desert patrols in North Africa, were engaged in chasing these smugglers.

After this excerise we spent a few days in Quetta resting, and our team playing in the finals of the Baluchistan soccer tournament, was narrowly defeated by the 8th. Gurkhas. Every evening of our stay in Quetta the officers foregathered in the Quetta Club where Harry Rich generously dispensed hospitality at the bar. My brother Louis invaraibly joined us, becoming an honorary member of our mess.

During March 1945 I heard about STIFF and SLIC, code names for two schemes relating to compassionate leave in Britain. SLIC allowed up to about a month for cases of mental exhaustion, serious family problems and the like. I could not in all fairness apply, but I found that STIFF allowed up to two months away to men who had been separated from relatives for very long periods. I immediately applied expecting to be high on the list, having now been away from England for almost ten years. As expected my leave was quickly granted and towards the end of April I set off for Karachi full of excited expectations at seeing my parents and brothers and sisters once more.

The war in Europe was in its last throes with the allied armies rushing across northern Europe towards Berlin. In Asia the battle for Kohima had been won by our 14th Army at great cost. and that fine force was starting to make its way South through Burma. The United States was island hopping in the Pacific, getting ever nearer to Japan, but meeting fanatical opposition at every turn. It looked as if the war in the east would continue for a long time, and surely the 5th.9th would be in it well before the end. I would like to enjoy my Home leave and return in time for the final act.

I arrived at RAF Lynham, Wiltshire, on 6th May 1945, feeling egg bound. Before boarding an RAF Transport Command, Dakota, at 4am on Maripur airfield, Karachi, our motley collection of miserable passengers were fed with egg sandwiches washed down by luke warm tea. I seemed to be the only STIFF member. The rest of my twenty odd companions came under SLIC, some being psychiatric cases, and others were hoping to sort out chronic family problems which had arisen during their absence abroad. They were not the most cheerful people to have along when I was in happy mood at the prospect of seeing my parents after ten years absence. Once in the air, sitting in the cramped canvas seats, we were handed egg sandwiches, and tea from thermos flasks. After about two hours flying we landed at Sharjar, in what is now Oman, where we were directed to the RAF mess for a breakfast of fried eggs. We took off again after refuelling, and were again served with egg sandwiches in the air. At about 11am, Karachi time, we landed at RAF Habbanaya, near Baghdad, to find that it was 9 am local time, and we were in time to partake of breakfast in the officers mess; eggs and bacon of course. We were told to do what we liked for the rest of the day, but we must be ready for take-off at 4.30am next day. The Dakota cabins were not pressurised, and their ceiling was insufficiently high for them to fly above the air turbulence caused by the heat over the desert. Hence the early starts, and the landings before the heat built up.Next day the performance was repeated; an egg breakfast in the mess followed by egg sandwiches in the air. We landed at Cairo during the morning and put our watches back another hour, in time to have the expected eggs for breakfast. I went into the city with the co-pilot, glad to get away from my gloomy co-passengers. We made the same early start next morning after the customary breakfast and, before landing at Tripoli, in Libya, ate egg sandwiches. From Tripoli we hopped across the Tyrrhenian sea and landed at a captured Italian air field in southern Sicily. We were taken to a beautifully situated ex Eyetye air force mess overlooking the sparkling blue waters of the bay near Cagliari, where next morning, for a change, our eggs were served with maccaroni.

The final hop over France and the English Channel was made on 6th. May. This time as we munched our egg sandwiches a few rounds of 'flak' exploded behind us. They were probably the last shots to be fired by the Channel Islands German Garrison. No breakfast was forthcoming when we landed at RAF Lynham but we were rushed off to Swindon's GWR station and I was instructed to report to the Station Hotel at St. Pancras. My first impressions, as seen from the train window after 10 years absence was of the amazing number of women pushing babies in prams,and the apparent normality of life in the countryside. After reporting and being allocated accommodation for the night, I made my way to Regent Street and Piccadilly Circus. Double Summer Time was in force, so it was still light at 1045pm, but the poor old place was battered, shabby and smoke-begrimed. Everywhere there seemed to be lounging GIs and touting tarts. Somehow I managed to contact Francis (Lt.Cmdr. F.H.Brooke-Smith G.C. RNR), and arranged to meet him next day which had been declared V.E.Day. Francis took me on a tour of almost every club in London, or so it seemed. He seemed to know everybody, and everyone knew him. I was still in tropical uniform which seemed to excite some interest. One girl,seeing my 9GR shoulder badges, enquired which railway I belonged to. Francis had arranged for us to put-up at a friends flat in Upper Berkeley Street, opposite the Cumberland Hotel The whole, of London was 'en fete', with bonfires in the parks and singing in the streets, and huge cheering masses outside Buckingham Palace, where the King and Queen accompanied Winston Churchill to wave from the balcony. The street lights were on for the first time since 1939. We eventually returned to the third floor flat to find it locked and no key available. Francis made an outside recce, noting some open windows on the same floor. The occupants of one of these flats were persuaded to let us pass through, and I followed Francis out of the window and along the narrow ledge, far above the cheering crowds, to enter our lodging by forcing the window. We were soon joined - by way of the lift and stairs - by about twenty of Francis' friends of both sexes. The party went on 'till after dawn,. After waking with a terrible hangover I realised I had missed.my train, and the next and the next, to Margate , where Louis'wife Jill and three year old Diana were to meet me. I had promised to make them my first call so that I could give them the latest news of Louis and his doings in India. They had been apart for about three years. I eventually got to Margate but the lunch, from valued rations, was spoiled. I sometimes wonder if I have ever been forgiven. Heartlessly I kept my parents waiting another-24 hours before going home to Suffolk. They made me so welcome and did not complain which made me feel very contrite. I think I was frightened at what I would find after so long. I need not have worried they seemed unchanged if a little older.

My two months passed too quickly. Brother Cuthbert was in northern Europe, commanding a Light Infantry battalion, sister Andrea was with the QAIMNS in Belgium, sister Margaret was with the WRNS and brother John was executive officer, in the training ship HMS "Conway", where youngest brother Guy was a cadet. I accompanied my parents on a visit to the "Conway", in the Menai Straits off Bangor. We went aboard the old three decker line-of-battle-ship, to attend Sunday Divisions and Divine Service ,taken by Captain Goddard. All the cadets were smart and the ship spotless. Then orders came to report to Thetford in Norfolk, to travel back to India with a large contingent of Ex POWs of the Indian Army. I was to be Adjutant and second-in-command of the draft, in the P & 0 liner "Corfu" converted for trooping, having recently served as a hospital ship.

At Thettord I was met by Lt.Col. Hays RIASC, a large, genial man who had been on the Indian Army Reserve of Officers prior to the war. He was commanding officer for the Indian drafts. We.went together to the local headquarters, situated in a stately home, which I have recently recognised as Euston Hall, the home of the Duke and Duchess of Grafton. The army had taken it over for the duration and happily it is now back in the hands of the owners who open it, and the the beautiful grounds, periodically to raise money for charities.

We were told that we were to take charge of some 40 VCOs and 1,400 other ranks of the Indian Army for the journey to Bombay, and would leave for Glasgow, the port of embarkation, in a few days time. I was delighted to find that the majority of the men were from two Gurkha regiments. They had been ordered to surrender, along with the rest of the Tobruk garrison in June 1942, after that town had been cut off from the rest of the Eighth Army by Rommel's Afrika Corps, and had bravely held out until ammunition and rations were exhausted. They had been held as prisoners of war by the Germans for almost three years.

These Gurkhas were immensly popular with the local Norfolk people, owing to their smart turn-out, impeccable behaviour and constantly cheerful grins. In the evenings the Gurkha Officers and men would visit the local hostleries. The GOs would fall the men in, by platoons, shortly before opening time then after a brief lecture about behaviour and the honour of their regiment, they would be marched off to the various pubs in the town. At closing time the senior GO or NCO at each pub would fall his men in outside. After checking his men he would call them to attention, report all present and correct to the publican, and request permission to move off back to camp. There was absolutely no rowdiness or bad behaviour, only cheerful laughter. All the ex-POWs had been carefully screened after release. The Gurkhas were passed one hundred per cent White. Of those from other units several had been classified Grey, and about a dozen were Black. The Blacks were known to have cooperated with the enemy, or had committed serious crimes such as murder or rape, and were held under close arrest. I suppose there were about 800 Gurkhas and 400 men from other classes in the Indian Army.

Lt. Col. Hays, having a great deal of experience in army administration, soon organised a well equipped troop train to take us all to Glasgow. I had the luxury of a sleeping coupe to myself and once we had got all the men on board, spent a comfortable night before arriving alongside the ship early next morning. It was a wet and murky day and the stevedores were on strike. When we started to load our own equipment they objected and demanded that we stopped. These objections were ignored and our men proceeded cheerfully to hump everything on board in record time. We found that in addition to our Indian Army contingent the "Corfu" was to transport about 80 British Officers, 58 VADS, 32 FANYs and 600 British Other Ranks. The VADs and FANYs ranked as officers so these ladies slightly outnumbered their male counterparts on the upper deck. The authorities had failed to organise suitable cooking facilities for the Indians and Gurkhas, not having understood the need for seperate cookhouses for Hindus and Muslims. The type of rations supplied were also unsuitable, and Colonel Hays had to use strong-arm tactics to get something approximating to Indian Army Rations loaded before the ship sailed.

As far as I was concernd the voyage was very pleasant since, being on duty, I was kept pretty busy. The ship was crowded but I had the privilege of sharing a good cabin with Hays. Most of the officers were jammed four or six to a small cabin, and the troops accommodation was in tiers of bunks in the holds. At this stage of the war troop ships were dry. That is dry for all troops being transported, but ships officers and those involved with the running of the ship were allowed a small daily ration of alcohol. I, being in the latter category,was allowed a bottle of beer and a tot of whisky daily.; to be consumed in the cabin.

We had scarcely left the Clyde before romances began to blossom between some of the VADs and FANYs and British Army Officers. This was not surprising as we were nearly all very young and the girls were leaving their homeland for the first time, going out to a strange country, and to be involved in war with a fierce and terrible enemy. Heavens knows when, if ever, they would return Home. It was an emotional time for them. The odds were against my getting involved at a time when I was pretty fully occupied with duties and all the others were free to carry on those shipboard romances. It was only during the last few days of the voyage that I was able to spend some time with a band of jolly VADS. The disembarkation orders, which I still have, gives no clues about them except that their draft was headed for Poona. They wore blue uniforms, so I assume they were VAD (RN). Perhaps they were on their way to Earl Mountbatten's Headquarters in Delhi or Kandy? I don't suppose I shall ever know. The three girls I remember called each other Tigger, Piglet and Eyeore! T.hey were very well behaved and very innocent.

When we left the Red Sea and entered the Indian Ocean we were met by typical South West Monsoon weather. It poured with rain, was very muggy in spite of the gale, and there were rough seas coming up on our starboard quarter. These made the ship pitch and roll and laid most of our passengers low, thinning the decks to a few hardy souls. On 6th August, when we were approaching Bombay, we heard the great news that a terrible new bomb had been dropped on Hiroshima, laying that city waste. This was the first atom bomb. Our reaction was to hope this terrible new weapon would lead to an unexpectedly early end of the war with Japan. We had understood that it might take years to defeat that country, and the Japanese would be expected to fight to the end, with terrible casualties inflicted on both sides.

There was great excitement on board as we steamed into Bombay Harbour and tied up at Ballard Pier. The British, particularly the girls, were moved by the anticipation of landing in a country reported to have mystery, splendour and poverty as well as great heat and wild animals, all so different to anything previously experienced. Our troops were returning to the lands of their birth, and to being reunited with their families after years of anxiety and absence., The excitement among the returning POWs was dampened considerably when an officer from the staff of the Bombay Area Command came on board and gave an address over the ships Tannoy, on behalf of his General. It was a short Welcome Home message, and was followed by the distribution, to each man, of a packet containing a   banana and a few Indian sweetmeats. Such was the welcome Home for soldiers who had bravely fought the enemy, and had nearly all withstood concerted attempts to suborn their loyalty. Disembarkation of the various drafts proceeded rapidly, with a succession of troop trains pulling away from the quay crowded with happy faces.

I stayed on board until the last of my ex POWs had been sent off to the transit camp at Kalyan, then I followed them. At 7.30 that morning I had handed over the 53 Blacks and Greys to an armed escort of 1 BO, 1 VCO and 15 men, for transit to Delhi. The Blacks, mostly Sikhs, had spent the voyage in the ships cells. I had noted the immense amount of military equipment and vehicles stacked on the harbour quays. It looked as as though preparations to send a large force from Bombay were nearing completion. This could only be connected with the recapture of Singapore and the Malayan peninsular.

The Kalyan Transit Camp was sited about 25 miles north east of Bombay. Access to the city was easy by frequent electric trains to Victoria Terminal. These trains ran with their wide sliding doors open so were pleasantly cool. Many passengers sat in the doorways with their legs swinging out over the rails. It was not uncommon for some of them to have their legs amputated by trains passing at speed on the opposite line, or even to fall off into the path of oncoming trains.. The weather continued to be soaking wet, muggy and depressing. I visited some of the "Corfu" drafts, now out of my charge and waiting at the camp for orders to rejoin their regimental depots. I found the camp's organisation deplorable. It looked as though the staff was running a lucrative racket. No mosquito nets were provided in a camp swarming with the insects. The barrack room lighting was supposed to be by kerosine oil lamps, but these were virtually unobtainable, so they spent the evenings in darkness. The rations were obviously of poor quality and insufficient quantity. I took this up at the office but was met with rudeness and obstruction.

On 9th August news came of the dropping of the second atom bomb, on Nagasaki. There was much speculation that the war might soon end.

During my enquiries at Kalyan, all of which I found objectionable and depressing, I came across a lone sick Gurkha rifleman, in a barrack room with some Madrassis. His story was amazing. He had escaped from the Japanese after being captured in Burma, or perhaps Malaya, I forget which. He had somehow made his way to Sumatra, then Java, and on through various islands of what was then known as The Dutch East Indies, to New Guinea. He was found by the Australians when they drove the Japanese out of the last named territory. The Aussies had shipped him back to India and he had been languishing in Kalyan for the past several months. He was a very sick man and should have been in hospital, but the organisation at the transit camp was such that the authorities probably were not aware of his presence. His journey was incredible under the conditions then prevailing and the man's sole aim was to keep out of the way of the Japs and to try to rejoin his unit.. I was still trying to get these matters sorted out when orders came for me to report to the 5/9th GR as soon as possible at Fort Sandeman. The same day, 14th August, we heard that Japan had capitulated unconditionally. I went into Bombay and celebrated before catching a train for the North. Every ship in the harbour was sounding its siren. Searchlights were piercing the low clouds and there were crowds in the Taj Hotel. I sent in a report about the situation at.Kalyan, and on my way to Fort

Sandeman left a copy with the Gurkha Depot at Quetta, and handed another copy to Harry Rich when I reached the battalion. Some months later I was delighted to learn that several of the staff at Kalyan had beencourt-martialled. Several severe sentences had been doled out and the sick rifleman had rejoined his unit. Actually things must have come to a head after I left as one of the charges was of mutiny.

During my absence on leave - much to everyone's disgust -, the battalion had been sent back to Fort Sandeman. I joined my Company in camp at Hassu Band, on the lower slopes of the 11,085 ft. Takht-I-Suleiman, about 25 miles from Fort Sandeman. It was a delight to be back in camp again with my men and we had a few days of marvellous shikar on the mountainside accompanied by B Company under Jim Fillingham. This camp was cut short by instructions to move to Wana in Waziristan. Wana was a frontier post in the tribal territory, very different from Fort Sandeman or Chaman. Tribal territory consisted of a mountainous strip of wild country lying between the boundary of Afghanistan and the administered districts of the North West Frontier Province. There was no law for the tribal inhabitants of this area apart from an understanding that they should not take pot-shots at our garrisons, and must refrain from raiding into the administered districts. These conditions were more honoured in the breach than in the observance. The garrison at Wana consisted of a Brigade Group; three infantry battalions, a battery of mountain artillery, plus animal and motor transport companies. The Camp could be compared to a Concentration Camp, being surrounded by the same barbed wire perimeter fence, and having watch towers and protected gates. The difference being that it was designed to keep the locals out and not us in. It was also compared to a monastery as no families were allowed. In fact it was said that no woman had ever been there. Adjoining the northern side of the perimeter there was a small air strip which was used occasionally for flying VIPs or urgent stores in or out. Nobody was allowed out of camp except in armed parties on duty. The road to India was opened on RODs (Road Opening Days) after the heights on either side of the route had been occupied by defensive piquets of our troops, or by the men of the local militia, the South Waziristzan Scouts. The ROD date was kept secret until the previous evening, but seemed often to be known in the camp bazaar before that.

The official history of the batallion states that;-

" With the surrender of Japan the last hope of active service disappeared and almost immediately, in the sad fashion inseparable from the ends of wars, good comrades began to disperse. Fifth Battalion had been exceptionally fortunate in retaining its officers for protracted periods; in three years only six new British Officers had been posted. The long months of waiting and of hoping had bred neither lassitude nor boredom; the Battalion had remained happy and offensive-minded in spite of its recurrent disappointments. It is now known that, as a matter of policy, it had been determined to retain one batallion of each Gurkha regiment in India throughout the war - a decision which perhaps was mercifully hidden from the unfortunate selections."

Bombay, Disembarkation details, 8th August 1945.

APPENDIX "L"

(Issued with Embarkation Headquarters, Bombay, Movement Control Disembarkation Order No. 131).

Copy No. _____

Vessel "81-Q". Berthing at 18 A.D. Date 8.8.45. Time

Disembarkation Staff:-

E.S.O. i/c Disembarkation.........Capt. H.J. KERRIDGE. R.E.
A.S.T.O.................................Lieut. BAGSHAW. R.N.V.R.
E.M.O.....................................Capt. C.P. THOMAS. I.A.M.C.
Embn Baggage Representative........Capt. WRIGHT. Bom Grens.

Disembarkation Date	Time	Draft	Unit	Os.	BORs.	VCOs.	IORs.	Destination.	Last meal on board.	Method of disposal.
9.8.45	0645 hrs. *0530*	RFJJN	Infantry.	2	160	–	–	NASIK ROAD.	Morning meal.	Depart by Special Train No.404/ Alexandra Docks at 0835 hours.
0400		RFJGG	Infantry.	3	246	–	"	"	"	"
		41746	54 Ind Rec Comp.	3	–	15	226	AURANGABAD.	"	"
		RGJFM	Pnr Corps.	2	–	–	–	JALNA.	"	"
9.8.45	0600 hrs.	RLVRM	V.A.D.	56	–	–	–	POONA.	Morning meal.	Depart by Train Service ox V.T
9.8.45 *0615*	0600 hrs.	RZANS *RNFGO BORO*	Ind Repats.	8	*29*	18	708	KALYAN.	Morning meal.	Depart by Special Train No.405/ Alexandra Docks at 1030 hours.
9.8.45 *0600*	0635 hrs.	RJFPO	R.E.	11	–	–	–	KIRKEE.	Morning meal.	Depart by Train Service ox Vict Terminus at 1100 hours.
		RJFQA	R.E.	21	–	–	–	"	"	"
		RZAQB	R.A.C.	2	–	–	–	"	"	"
9.8.45. *0600*	0715 hrs.	–	Ex P.O.W.	–	–	–	53	DELHI.	Morning meal.	Depart by Train Service ox Bomb Central at 1810 hours.
			Note:- Escort of 1 Off, 1 VCO & 15 IORs will accompany the above POWs.							
9.8.45.	1600 hrs.	RMFKZ	R.Signals.	2	–	–	–	MHOW.	Midday meal.	Depart by Train Service ox Bomb Central at 1845 hours.
		RMFGO	R.Signals.	29	–	–	–	"	"	"
9.8.45. *0745*	1700 hrs. *0600*	RZANJ	Ind Repats.	6	–	9	462	KALYAN.	Evening meal.	Depart by Special Train No.406/ Alexandra Docks at 1900 hours.
		RGJOA	R.E.M.E.	11	–	–	–	"	"	"
		RPJLK	Various.	8	–	–	–	"	"	"
		RGJOO	R.E.M.E.	–	5	–	–	"	"	"
		RGJRA	R.E.M.E.	–	17	–	–	"	"	"
		RRKGG	R.A.M.C.	1	–	–	–	"	"	"
		RJFPO	R.E.(Tn).	–	95	–	–	"	"	"
		RJFQA	R.E.	–	45	–	–	"	"	"
		RJFPO	R.E.(Survey).	–	22	–	–	"	"	"
		RJFQA	R.E.	–	14	–	–	"	"	"

One meal advance for 3 days at about 1430

Continued on sheet two.

Disembarkation details (continued)

Shoot Two.

Disembarkation Date	Time	Draft	Unit	Os.	BORs.	VCOs.	IORs.	Destination	Last meal on board.	Method of disposal.
9.8.45	1700 hrs.	RUBPV	Int Corps.	1	10	-	-	KALYAN.	Evening meal.	Depart by Special Train No.406/A a Alexandra Docks at 1900 hours.
		RUBPV	Int Corps.	-	15	-	-	"	"	"
		RZALO	French Forces.	10	16	-	-	"	"	"
		RJFPN	R.E.	-	3	-	-	"	"	"
9.8.45	#900 hrs. 0840	RJFPO	R.E.	-	3	-	4	JUBBULPORE.	Evening meal.	Depart by Train Service ex V.T. a
-	-	RZAJR	F.A.N.Y.	30	-	-	-		-	Date and time of disembarkation to notified later.
		-	War Office Civilians.	2	-	-	-		-	

B A G G A G E   P A R T I E S.

Draft.	Os.	BORs.	VCOs.	IORs.	Remarks.
41746	-	-	1	2	
RZANS	1	-	1	8	
RZANJ	1	-	1	6.	
ROJOA	1	-	-	-	
RJFPO	1	10	-	-	
RPJLK	1	1	-	-	With nominal roll.
RJFPN	-	1	-	-	
RMFKZ	1	1	-	-	
ROJRA	-	3	-	-	
RPJJN	1	6	-	-	Will also look after kit of ROJOO.
RPJGG	1	8	-	-	
RUBPV	1	2	-	-	
RJFQA	1	6	-	-	
RZAQB	1	-	-	-	
RGJPM	1	-	-	-	
RRKGG	2	-	-	-	With nominal roll.
RMFGO	3	-	-	-	" " "
RLVRM	3	-	-	-	" " "
RZAJR	1	2	-	-	" " "
RZALO					

all people for 0800
0380 hrs each tr...

Sweepers & rem...
found until o fo...

CHAPTER ELEVEN

WAZIRISTAN - DEMOBILISATION - RETURN TO CEYLON

"The Almighty created in the Gurkha an ideal infantryman .... brave, tough, patient, adapt able .... intensly proud of his military record and unswerving loyalty. Add to this his honesty in word and deed, his parade perfection, and his unquenchable cheerfulness, then service with Gurkhas is for any soldier an immense satisfaction'

Field-Marshal Lord Slim.

The first officer to leave us was W.A. (Bill) Thom, our tough Rhodesian. His wife had arrived in India but was forbidden to accompany him to our Frontier station in tribal territory. He was followed by R.E. (Bob) Bailey our "Old China Hand". They were replaced by Stannard, Kinnear and Gill. All three newcomers were good keen men. Harry Rich's hard training had achieved its purpose and the battalion was now enthusiastic and confident. Although the war in the East, against the Japanese, was officially over there would be much mopping up and sorting out to be done. This actually took years to accomplish and proved a depressing and heart-rending business for those involved. It seemed that we would be spared from that unpleasant duty. Instead we would be guarding the real tribal North West Frontier, a very different matter from policing Baluchistan. The politicians were negotiating terms for India's independence and this gave rise to much speculation and many rumours. The whole Frontier appeared to be waiting with bated breath to see what Britain would do. Would there be internal strife affording the tribes the opportunity of raiding and looting in the plains? Would the military have to withdraw forces from the Frontier to cope with unrest in India? Would the mullahs try to inflame the tribesmen to fight us, as in 1919 after the First Great War? These were some of the possibilities we were ready to face.

Arguments had started among the politicians, who knew little or nothing about the matter, about the future of the Gurkha Brigade, whose officer cadre - from 2nd. Lieutenant upwards was British. These officers, like the Gurkhas who served them, often followed, generation after generation, from grandfather and father to son. This was still true of the men, even with the immense war-time expansion. Thus each regiment was something like an extended family, and had remained so for over one hundred and twenty five years. I was proud to be serving in a regiment with such traditions, and hoped there would be a great future for the unit. I was in no hurry to be demobilised. In any case it seemed likely my demob papers would be well down the list, and I hoped a decision about the Brigades' future would be reached long before my number came up. Meantime I would keep my eyes and ears open and hope for an opportunity to apply for a regular commission in the 9th. Gurkha rifles.

Senior Gurkha Officers of 'A' Company, 1945

Jemadar Kharakbahadur Mall,
Pl. Cmdr: A. Coy, 19 years service.

Subedar Ganjbahadur Gharti,
Pl. Cmdr: A. Coy, 21 years service.

Subedar Tekbahadur Khattri (2i/c),
A. Coy, 23 years service.

We were soon sent out on RP - road protection - This was a hard task and we did it often. As often as the road had to be opened for supplies to reach our brigade in Wana. We could not risk letting our convoys travel the roads under the questionable protection of the khassadars alone. On RP days we got up early and marched out in battle formation searching every place where an ambush could lie in wait. As we reached every vantage point we dropped off groups of soldiers called piquets. At the end of our sector another battalion took over and continued the process or we met troops from the next camp who had been working out towards us from their end. When the whole route was protected signals were sent to base that the road was clear and the waiting convoys at each terminal immediately set off. We covered the same stretch of road on every RP day and this made it difficult to obey cardinal role of the Frontier, viz, "never do the same thing in the same way twice running". Someone was always watching. Someone who missed nothing. Someone with eyes like a hawk, a born tatical sense, and plent of patience to wait for a suitable opportunity.

The background to all this was that the British had, ever since 1846, been faced with the task of trying to pacify the Pathan tribes living along the border between Afghanistan and India. they never entirely succeeded. Most of this area became known as the North West Frontier. During the 3rd Afghan War of 1919 when the tribes joined in the invasion by the Afghan forces large number of Wazirs and Mahsuds had deserted from the frontier militias, taking their arms with them. So many Afridis of the Khyber Rifles deserted that the unit was disbanded. During the following few years hundreds of raids were launched on towns such as Peshawar, Kohat, Bannu and Dera Ismail Khan, causing the deaths of hundreds of British subjects, the wounding of many more plus over 450 kidnapped.

In the hope of getting a better grip of the situation the British decided to advance into tribal territory and set up garrisions at strategic places such as Razmak and Wana. However this move only succeeded after some of the fiercest fighting yet seen on the Frontier. Wana was located in the heart of the homelands of the fiercest of the frontier tribes, the Wazirs and the Mahsuds. The place had a bloody history. Its rocks, cliffs, and stony plateau kept an eloquent silence on its continuous record of bloodshed. In this bleak upland, girdled by rugged hills, the camp was surrounded by a stout perimeter wall, the watchtowers manned at all times by alert armed sentries. Part of the brigade was always on stand by for immediate call out. The hills overlooking the camp were occupied by armed piquets as an additional precaution against raids. This extraordinarily rugged country has produced the toughest and most warlike people on earth. There is a tale told by the tribesman that when God created the world he found that a great many stones and rocks were left over which he decided to pile in heaps along the Frontier. Everything is hard and bare, toothy crags bite the sky, and in winter the cold and wind off the Hindu Kush are benumbing, whilst in summer the heat becomes unbearable.

The Wazir has been described as like a panther and the Mahsud as a wolf. Both are splendid animals. The panther is sleeker and more graceful while the wolf pack is more purposeful and hunts as a team and therefore is more dangerous.

This is the country, and these are the people, of the territory around Wana. Soon after the difficult first occupation of the place by a brigade of all arms it was almost overrun. The tribesmen had waited and watched for many nights until, at last, they saw their opportunity. Just before the break of day their silent lightning rush carried them right to the middle of the camp, hacking, slashing and shooting. Fortunately the gunners were alert and quickly turned their guns inwards, firing into the mass of the attackers who melted away almost as quickly as they had come. Later on the camp commandant was brutally murdered with a bayonet thrust through the heart. The slayer of another officer claimed he fired the shots "because the sahib slept with his feet pointing towards Mecca". Our arrival had been preceded , just a few years before, by the ambush in the Shahur-Tangi, on the Manzai to Wana road, described on another page.

All Pathans are constantly in need of arms and ammunition. The easiest way to get these was to shoot a soldier out on manoevres. There was a fascination in the atmosphere of a brigade column winding its way along the valley with the heights being piqueted on either side. The soldiers hoped there would be action and shooting but the Political Officers hoped there would not. The little band of POs liked to run the show alone. They claimed the army was only there to come to their aid when all other means of keeping the peace failed. The POs carried on their very dangerous task at constant risk to their lives. Surprisingly, perhaps, many grew to love this barren land and its warlike people. To remain alive it was essential that they were constantly vigilant and knew and understood intimately, the local dialects and customs. This may be illustrated by just one example, The story was that after some particularly vicious banditry by a section of the Mahsud tribe the PO called a jirga and opened proceedings by telling them what the penalties would be, including the fine. This was badly received and there was much barracking. The PO went on to demand that the leaders hand over all the main culprits. This caused tumultous objections. "That's your job. It's up to you to catch them. Aren't you the PO? Isn't that what you are paid for", etc? The situation was becoming electrically tense when there came a thunderous bellow from the back; "You are the PO. Haven't you got any balls"? It was a freezing morning. The PO thrust both hands into his trouser pockets, groped around, and shouted back; "well, I thought I had, but it's so bloody cold that I can't find them". There was a great roar of laughter, and the Jirga dispersed in high good humour, promising to do anything he wanted. One such Political Officer was worth an army brigade and cost only a tiny fraction of such a formation. However the army had to be there, in case!

Just now, however, the Frontier seemed peaceful. But it would be dangerous to relax because the tribesmen always played a game of waiting and watching. Furthermore the political situation at the end of the Second World War seemed rather like that after the first, in 1919, from which many of these troubles arose.

I wrote Home in a letter to my parents in September 1945: "The Pathan is a wonderful soldier. He lays an ambush and always waits patiently until he sees a mistake. He never misses one, or a weak spot. Then he attacks. If we take adequate precautions and make no mistakes, he remains in position, perfectly camouflaged, and invisible,until we have passed by, or else sneaks away down some side ravine, unseen by us. When we get a casualty from sniping we seldom see or locate the sniper. It is an absolute rule, on the Frontier, never to leave a Casualty. He MUST be recovered, and so must his weapons. We are expected to retrieve our spent ammunition cases too. A unit is judged by this. If the tribesmen get hold of a wounded or dead man they inflict frightful, unspeakable, mutilation, and weapons and used ammunition cases are used against us (after refilling the cases), to cause more casualties. At the moment things are quiet, although recently we had regular sniping into the camp at night. The brigadier suspected the culprits were khassadars, tribal police who are supposed to cooperate with us. So he issued them with tracer ammunition. Next night tracer bullets winged their way into camp so he proved his point and the sniping has ceased, for the time being. We had the last laugh that time.
P.S. I have taken over a very nice Staffordshire Bull Terrier pup from Bill Thom who has returned to Rhodesia as a Lt.Colonel."

A Pathan of the Mahsud tribe.

The Khassadars were much used by the British Political Officers, but their unreliability was well known. They often formed a "bedragga", a bodyguard, when a PO travelled in tribal territory. The PO would sometimes insist that the men sit on the roof of his vehicle since this tended to dissuade tribesmen from taking pot-shots at him for fear of hitting a khassadar and starting a blood feud. Patrick Duncan, one of the few political officers who stayed on the Frontier after independence was shot by a Mahsud khassadar whose pride got the better of him at a jirga. Duncan did not deal with his request at once as he was busy talking to other Maliks, and the tribesman, feeling insulted by this apparent dismissal, shot him.

While we were at Wana in 1946 the Political Agent, J.O.S. Donald, was kidnapped due to the failure of his bedragga of khassadars to resist. The road was blocked and the tyres of his lorry were shot out. After ten days he was released on payment of a large ransom. Such action Could not go unpunished, and the Wana Brigade was sent out to deal with the responsible Shabi Khel people. Unfortunately Donald became very depressed and shot himself. He apparently felt he had been paying out allowances to maliks and khassadars only to find they could plot against him and desert their posts at the drop of a hat. The bedragga system had failed, and all he had striven for was in vain. The death of Donald was a sign of the toll which could be taken by the strains of Frontier life. During the autumn of 1945 I went on another course, to Dehra Dun. On the first part of the return journey to Wana, on the train from Delhi to Lahore, I shared a coupe with an officer from another

Gurkha regiment and our Gurkha orderlies shared the servants compartment at the end of our coach. We told them to wake us with chota hazri at Amritsar, about 30 minutes before reaching Lahore. We both woke to find the train stopped and on looking at my watch I noted we were overdue at Lahore station. We were in fact stopped in that station and had to dress quickly and get off the train with our bedding rolls and other gear. There was no sign of the two orderlies and their compartment was empty. We rushed to the RTO's office and got him to search the train before it left for Rawalpindi. Telephone inquiries to Amritsar proved negative. After a discussion it was decided that my companion would wait at Lahore to check trains arriving from the East while I returned on the next train to Amritsar, in case the men were stranded there. I drew a blank at Amritsar and a telephone call to Lahore revealed that my friend had not had any luck his end.. When I got back to Lahore it was getting late so I decided to stay the night there. Later the same day a message came in that two Gurkhas had turned up at a small wayside station,down the line, and had forced the station master at gunpoint, to stop the Frontier Mail, India's most famous express, to pick them up at the little halt. A couple of hours later when the great train rolled into Lahore station, two bedraggled little Gurkhas descended to the platform and, as we approached them, drew themselves up, shouldered their rifles and brought their left arms smartly across their chests in salute. Both men were in pretty poor condition and Bhimbahadur, my orderly, was covered in blood and obviously far from well.

They told us that strangers had entered their carriage during the night and seemed to have drugged them after friendly overtures with food and cigarettes. They remembered a fight and the Budmashes trying to sieze their weapons before they were pushed out of the speeding train. They fell onto a shrubby bank and, after picking themselves up, walked along the line to the nearest station, a tiny halt, where they told their tale to the station master. He mentioned the next train up the line would be the prestigious Frontier Mail. The Gurkhas told the SM he should stop that train so that they could rejoin their sahebs as soon as possible. It was their duty to do that. The SM declined but eventually, under threat of the dire consequences of refusal, he gave way and flagged down the great train. I could well imagine the fear of death they put into the SM.

It was obvious that this incident would cause quite a stir and from Bhimbahadur's appearance I could tell he was in need of immediate medical attention. The other orderly was fit to travel and left with his officer for Sialkot. A doctor found that Bhimbahadur had broken several ribs, fractured a bone in his foot, plus various lacerations and possible internal injuries. It was remarkable he had been able to walk at all. I left Bhimbahadur in a military hospital in Lahore where he remained for several months and although I wrote a report indicating that, in my view, the men had not only behaved blamelessly but with considerable initiative in carrying out their duty to rejoin their officers, a court of enquiry was not held until after my demobilisation and departure from India. I never heard the outcome in spite of several enquiries. By that time the batallion, now the 2nd. 9th., had moved to the Punjab where companies were widely scattered to deal with the mounting chaos and civil disorder which preceded the division of the old India.

Generally conditions on the Indian Railways had by this time become pretty bad, but of course nothing like as bad as they were to get later during the partition period. The worst I experienced was when twelve officers had to share a two-berth coupe. A very junior man whined about the conditions and took over the top bunk for himself. The rest of us, including a Brigadier just back from the Burma battles, spread our gear on the floor and slept head-to-heel across the carriage. It was staggeringly hot though!

Wancol, October 1945

Waziri tribesmen near Wana, 1945.

3.5 inch mountain gun in action.

Carriers near Tarnai, Waziristan, part of escort to 5/2 GR on move to Damdil.

Wana brigade, below Michin, Baba.

My brother Cuthbert (Tup) came out to India, to attend the Staff College at Quetta, late in 1945. I was unable to get to Quetta to meet him. According to a letter I sent my parents, "I was unable to get through (to Quetta) because the intended Road Opening day between here and Fort Sandeman on 24th December was cancelled owing to trouble on the road. Anyway I would have had to come back the long way round (thousands of miles around three sides of a rectangle), which would have meant two days in Quetta and eight days travelling, with considerable expense. In any case having been Home this year I am eligible for very little leave." Tup was a Lt. Colonel, having commanded a battalion of Light Infantry during the battles in northern France. My sister Andrea also came to India with a detatchment of QAIMNS. I wrote to her at Port Said suggesting she try to get posted to Quetta, Rawalpindi or Peshawar, as healthy spots with good hospitals and accommodation. My letter missed her and she was posted to a hospital in Assam at the other end of the subcontinent. Louis had returned to Britain but Francis was at Hong Kong, so there were four of the family east of Suez

Instead of going to Quetta I had an enjoyable Christmas shooting on the banks of the Indus near Dera Ismail Khan with Jim Fillingham and a section of eight Gurkhas. I was convoy commander for the Road Opening between Wana and Manzai, the rail-head sixty miles to the east. This was a fairly relaxed affair as the route had been picquetted and all I had to do was to get the vehicles through as speedily as possible. Fortunately there were no delays or breakdowns to complicate matters. We stayed a night at Manzai and I wrote that, "Manzai is a terrible place situated on a long bare ridge and there isn't a tree or a blade of grass for about twenty miles in every direction. The "Heat Stroke Express" runs from here to Tank, then the railway heads north to Attock, but we went east to the Indus at DIK,in our own trucks and put up at the dak bungalow for Christmas day and that night. DIK is an old Frontier station which has been in existence since before 1850 and, being near the Indus, has plenty of trees and cool old bungalows. What is best is that it is not surrounded by barbed wire and you can get outside whenever you want. The Gurkhas came to act as beaters and general helpers around camp, which they love doing. We chose especially good men as it is a great change for them from the boring routine here. At first they couldnt realise they need not have rifles constantly loaded at the ready, or to keep a continuous look out.

Early on Boxing Day we got hold of a boat, a large flat-bottomed native boat about thirtyfeet in length and fourteen feet beam. We put our tents and kit on board and floated off down stream, propelled by the current. The river consists of many small channels varying from 100 to 600 yards wide, and in all about 14 miles across from the extreme western to the extreme eastern banks. We kept near the eastern side, where between channels is low land covered by long coarse grass 6 to 7 feet high, in which it is easy to get lost. In amongst the grass and scrub there is plenty of game, particularly wild boar. The boat cost us Rs.2 a day. After floating downstream all day, occasionally punting off sandbanks, we pitched camp on a suitable mound in the evening and had a meal of curried duck. We also shot a few partridge near the camp. On the 27th, the boatmen started the slow pull up stream towards DIK while Fill, the Gurkhas and I swept along the bank on foot. A flock of several hundred geese came over and settled on a mud flat, but their sentries were too alert for us to get within killing range. The 28th was a more successful day with Fill getting a fine black buck. The men and boat coolies were as happy as sandboys with the plentiful meat. On 29th we had great fun going after wild boar. It was exciting walking them up in the long grass. Some were huge, more like rhinoceros than pig. Next evening we arrived back at DIK and put up at the Mess of an Indian cavalry regiment. Whilst being most hospitable the cavalrymen were rather scathing at us going after boar on foot with shot guns. Apparently that is not at all the thing to do. The only decent way is to go after them with lances on horseback. "Stick-em dear fellow! Stick em!".

Buscol 'Stand-To' Company, 5/9th Gurkha Rifles, near Tarnai, Waziristan, February 1946.

On our return journey we were held up for a couple of nights at Manzai. The road to Wana remained closed due to reports of tribesmen gathering near the Sahoor Tangi, a wild and barren pass with sheer cliffs on each side. Scout posts at Lodda and Sararoga were also said to be cut off. The PA did not want any trouble which would mean having to send out a punitive expedition in force. For various reasons this would not be a good thing just now, whilst the internal situation in India is so shakey. So he just ordered that no convoys go through. Maybe he thought that in this very cold weather the tribesmen would not linger anyway. On the 2nd there were reports that the tribesmen had cleared off, so on. the 3rd. we went through without any shooting at all. But the wildness of the Sahoor Tangi has to be seen to be believed. In 1938 a convoy was ambushed there. 8 officers and 73 troops were killed and over 100 wounded, so they take no chances. Tangi means "tight place" and the Sahoor is very narrow and steep. During the ambush the armoured cars were unable to elevate their guns to get at the tribesmen firing down from the cliffs. No doubt they had worked it all out before-hand. The evening after we got back there was sniping during a game of football outside the wire, and one spectator was wounded.

On the 6th I took a new horse, one of the chargers from the mountain gunners, which is a re-mount, and only half trained, and rode out with a relief picquet to see the changeover. I had a lot of trouble with my mount. He reared and bucked but I eventually got him going alright. I checked the picquet and started back with the relieved garrison, when he started his tricks again, rearing up with his forelegs right up in the air. I slipped in the saddle bringing a lot of weight onto the reins, and he.went right up and over onto his back, falling on top of me. Then he got up and bolted. The men caught him,and brought us both in. I thought I had broken a leg and an arm, but the doctor said I was very lucky and had got off with a severe bruising of my leg and only a slight fracture of a bone in my left elbow. So I am now limping around with my arm in a sling and feeling a right old charley. I have heard no more about my release, but if the firm's application doesn't succeed I hope to be here for some time yet. It's all worked out by the Man Power Board in Ceylon".

I was not aware then that Jock Sutherland, the Manager of the Spring Valley company, and his No.2, Jack Cranfield, had been agitating for some time for my return under the Class B Release Scheme. Old Colonel Maxwell Johnstone, who had taken on my job at May Mallay in January 1942 had already left Spring Valley which was now very short-handed. Both my seniors on the estate were long overdue for Home Furlough and had been running the company as well as overlooking other estates whose superintendents were away on war service. Managing extra properties under war time conditions with shortages of food and materials and relying on rather unskilled subordinates caused considerable strain, and they and their wives health was beginning to suffer. Thus I was in a quandary; as long as the future of the Gurkhas was undecided there would be no promises of regular commissions in The Brigade; whilst should I decline release under Class B my firm would most likely fill the vacancy with somebody else and I might find myself out of work. I decided to keep a low profile and hope events would run in my favour.

Meantime life in Wana carried on in much the same way except that Harry Rich was transferred. Pending the arrival of a new commanding officer Peter Richardson officiated as commandant and I as his second-in-command. I even commanded the batallion for a brief spell as can be seen from a sample of one of Battalion Orders issued in my name illustrated on another page. As the unit demob officer I was responsible for completing all the arrangements for sending home the pensioners who had so speedily re-joined the colours at the outbreak of war. These were followed by a few men from the ranks but most of our soldiers were happy to continue their service as long as they were able to get a spell of leave to see their relatives in the mountains of Nepal. Then orders for my own release came through on 11th May 1946. No announcement had been made about the Gurkha Brigade so, perforce, I set about organising my own demobilisation. I struck myself off the strength on 2nd. June which happened to be the day following the reconstitution of the 5th. Battalion on the 2nd. Battalion. However before those events came to pass there was one last column. Peter Richardson commanded, I was his 2i/c and Jim Fillingham of B Company was the only British Officer company commander. The other three rifle companies were commanded by Gurkhas.

The Brigade had carried out a ten day sweep in November along the Afghan Frontier as far south as Toi Khulla. That exercise had passed off with relatively few incidents, though a few Wazir snipers lay among the hills. That perennial malcontent, the Faqir of Ipi, had transferred his attentions to the north where, with newly acquired artillery, he shelled Razmak sporadically. That the countryside was not as empty as it seemed was demonstrated on the last evening before return to Wana, when a fusillade from a near-by ridge at evening stand-to resulted in the death of a muleteer. The column of April 1946 came about as a result of the kidnapping of the South Waziristan Political Agent already mentioned. Our force consisted of Fifth Battalion as Advance Guard with 7 Baluch in reserve. 4 Dogras was garrisoning Kharab Kot Camp, and we passed through 1 Ajmer who had earlier moved out from Wana to picquet the adjoining hills. In support of 5th. Bn. we had one battery 9th Mountain Regt. 3.5 Howitzers and a section of Field Ambulance. In addition to the mules normally attached to each unit some 250 baggage mules were to follow through. We were spared the customary lines of shambling camel pack bearers on this column. At Dargai Oba camp the tribesmen were in playful mood and managed to steal all our latrine screens.

On the sixth day after a cross-country march to the west, we were given the task of picqueting Tora Tizai, a dominating feature surrounded by very broken ground with shrubs covering the nullah banks, affording excellent cover. C. Company was sent up peak 5333 about half a mile west of Tora Tizai, accompanied by an FOO. Artillery was trained on the peak, while C.Coy.under Subedar Dammarbahadur leap-frogged his three platoons across the broken ground. At about 1050 hours rifle fire was heard from the direction of Pt.5333, followed by a burst of LMG. At battalion H.Q. it was not possible to see details of what was happening and calls over w/t (by now issued to each platoon), revealed that in the stiff climb the operator with the heavy set had lagged behind, so it was several minutes before we learned that two of our men had been hit. Peter then ordered artillery fire on rocks a few hundred yards south east of 5333. The first shot was bang on and he ordered five rounds gunfire. As second-in-command I seemed 'de trop' and, to get away from the flap and Brig. Groves, I asked if I might go forward with my orderly and spare w/t. Having received permission we crossed the very rough ground of the lower slopes while farther ahead and higher up we could see some C.Coy men moving to the right and heard more firing from the same direction. We arrived at Coy HQ to find Sub.Dammarbahadur in calm control and soon after a wounded lance naik and a rifleman were carried in. The artillery again plastered its target and a few minutes later the picquet signalled it had reached its objective without further opposition. Feeling that by remaining with C Coy I would only be in the way, I returned with the wounded men to Bn. HQ where the casualties were handed over to the ambulance unit.

When all the troops and transport had passed through there remained the tricky business of withdrawing the picquet and the rest of C. Coy across the broken ground and back to camp. The picquet had about 700ft. of steep hillside to descend to reach the rest of the company. This was broken by fissures and small cliffs in which boulders and small bushes were intermixed. After this murderous incline they had to cross a narrow, steep sided nullah, pass through the other platoons of C.Coy., and run across several hundred yards of broken plain to reach rearguard HQ, now taken over by us. The rest of C.Coy. would cover the initial withdrawal as far as the nullah (dry stream bed), and then in their turn leap-frog back and past us.

MESS BILL.

Dr.

NAME *Major C.F. Brooke Smith*

To **Officers' Mess, 5/9th Gurkha Rifles.**

For the month of ___*May*___ 1946.

CHARGES.		Rs.	As.	Ps.
Mess Subscriptions	...	8	-	-
~~Servants~~ Wages *Hewing & lighting*	...	5	10	-
Messing	...	100	12	-
Maintenance Charges	...	30	-	-
~~Fines~~ and Cigarettes	...	149	-	11
2% Profit ~~on Wines and Cigarettes~~ *Entertainments Fund*	...	3	-	-
B. ~~Or Purchases~~ *Agency (Details on Reverse)*	...	25	7	-
~~Stamps and Cash~~ *Ice.*	...	3	-	-
Mess Paper and Periodicals	...	12	-	-
Wireless	...			
Mess Guests	...			
AMOUNT DUE TO MESS	...	334	1	-
ARREARS :—	...			
CREDIT :—	...	260	-	-
NETT AMOUNT DUE		68	6	-

J.D. Kumar
Mess Secretary, 5th Bn. 9th Gurkhas.

Dated 1 *July* 1946.

Place Wana

NOTE :—1. Please add a/. %..... exchange on outstation cheque.

2. Cheque should be made payable to Mess Secretary, 9th Gurkha Rifles and should be crossed. Payment should be made by the 15th instant vide R.A.I.

3. Officers are requested not to deduct any sums from this bill. If there is any error, please point it out and it will be adjusted.

Albert (Elec) Press, Quetta.

RESTRICTED
BATTALION ORDERS
PART I
BY
MAJOR C.F. BROOKE SMITH
OFFG. COMMANDANT. 5TH BN 9TH GURKHA RIFLES

WANA.
Dated 26.11.45
Dated 21.11.45

5/9 G.R.
Serial No.73
Last Part I Orders issued : Serial No.72

DUTIES.

1.
   G.O. of the day for 26.11.45 Jem Khembu Mall HQ.
   G.O. of the day for 27.11.45 Jem Dharamraj Khattri A.
   G.O. of the day for 28.11.45 Sub Tikaram Khattri C.
   G.O. of the day for 29.11.45 Sub Damnarbahadur Mall D.
   G.O. of the day for 30.11.45 Sub Bhadrabahadur Mall B.

   Gd.Comdr on 19.11.45 No.102979 Nk Lachhuman C.Coy
   Turn Out - Good   Drill - Good
        Cleanest man - 101102 Rfn Lilbahadur
   Gd.Comdr on 20.11.45 No.2532 Nk Rambahadur C.Coy
   Turn Out - Good   Drill - Fair
        Cleanest man - 103809 Rfn Ranbahadur
   Gd.Comdr on 21.11.45 No.95989 L/Hav Balbahadur C.Coy
   Turn Out - Fair   Drill - Fair
        Cleanest man - 97441 Rfn Pirthbahadur
   Gd.Comdr on 22.11.45 No.96699 Nk Jagatbahadur C.Coy
   Turn Out - Good   Drill - Good
        Cleanest man - 97444 Rfn Chitrabahadur
   Gd.Comdr on 23.11.45 No.102979 Nk Lachhuman C.Coy
   Turn Out - Good   Drill - Good
        Cleanest man - 98098 Rfn Lilbahadur
   Gd.Comdr on 24.11.45 No.2532 Nk Ranbahadur C.Coy
   Turn Out - Good   Drill - Good
        Cleanest man - 95634 Nk Ranbahadur D.Coy
   Gd.Comdr on 25.11.45 No.95634 Nk Ranbahadur D.Coy
   Turn Out - Good   Drill - Fair
        Cleanest man - 97312 Rfn Indrabahadur Khandka.

ORDNANCE - AMMUNITION. Bn Part I Orders Serial No. 46/7/45 is
2. republished below:- I.A.O.725/45 is reproduced.

   Ammunition - Necessity for care when returning live ammunition and ammunition empties to ordnance.

   Many complaints have been received that ammunition and ammunition empties have been returned in a dangerous condition by units to ordnance ammunition establishments. It is the responsibility of unit commanders that ammunition and explosives in any form are returned to ordnance properly packed and the packages correctly marked.

   2. Units will NOT :-

   (a) return grenades and mines anti-tank primed or fuzed or deficient of safety components such as tapes, safety bolts, pins, collars, Caps, etc.

   (b) return blind projectiles, grenades, mortar bombs, fragments or components containing explosives which may be recovered after partial detonation.

   (c) return live or misfired S.A.A.mixed with fired cases.

   (d) return Q.F. cartridges fitted with misfired primers.

        Misfired primers will be packed in a separate

3. ...lishments should be packed in its original package and container. All live ammunition returned to Ordnance esta- Where this is not possible units will use packages which will ensure maximum safety during transport and handling. All packages containing ammunition returned by units will invariably be marked clearly giving details of the contents.

4.

158

After the "Prepare to Retire" signal had been flagged we could see the first pairs of men start down to commence the thinning out process. The heavy wireless which would have held up the final withdrawal, and some of the slower men, or men with minor injuries, were the first to go. Suddenly there was firing from the remainder of the platoon on the hill-top, and a message to say tribesmen had been seen in the dip close to the north. Peter gave the order. "No. 6 picquet Retire", which was quickly flagged by the keen little signaller standing at his elbow, i.e. RTR. As the last dot of the signal was acknowledged the picquet began to move - the orange screens on the chests of the two flank men were clearly visible - and the whole lot plunged down the cliff like falling boulders. They descended in great crashing bounds, leaping six to eight feet down at a time, leaning forward and their legs working as fast,as pistons. It was only seconds before they reached the nullah and passed through the rest of C.Coy. The speed at which they had moved was unbelievable and gave the tribesmen no chance to get near them. Only Gurkhas are capable of descending at such speed. As soon as the picquet was clear of the top a torrent of 3.5" howitzer shells churned the earth at the point they had just left.

The picquet was now running strongly across the comparatively open ground, then as they reached us, the men closed up and grinning broadly trotted past, followed by the rest of C.Coy. by platoons, with Sub Dammarbahadur rolling along in the rear with his orderly and signaller, waving his stick and encouraging his men in his harsh rasping voice. C. Company had done its task perfectly.

I have little recollection of the rest of the column, except that when we got to Shabi Khel country the birds had flown and the engineers destroyed the tower houses unhindered before we returned to Wana.

Our return was greeted with the news that Second Battalion would be reconstituted on Fifth Battalion. After the recapture of Singapore by the British the survivors of Second Battalion had been returned to India and, after a period of recuperation at Dehra Dun;they had all had leave in their Nepal villages. Many of the men were ill or worn out by their ordeal during the years of captivity and, after return from their leave, it was found that only 250 of the old battalion were fit for further service. Fifth Battalion therefore supplied over three quarters of both officers and men for the new unit. Meanwhile Lt. Colonel C.S. (Sidney) de Wilton took over as commanding officer. de Wilton was a charming man with much humour but lacked the fire and character of Harry Rich. My departure from Wana followed immediately after the ceremony of the reconstitution of the Second Battalion which involved an impressive parade when the old Second Battalion men trooped through the ranks of the Fifth. This was followed by the lowering of the 5th. Bn's flag and raising of the Second's, to the bugle calls of the Last Post followed by Reveille. The new battalion then marched past Brig. Groves with pipes and drums playing and colours flying. This was followed by a sports meet and nautch to celebrate the occasion. So my last memories of service with the Ninth Gurkha Rifles Are of a warm and friendly atmosphere, rum flowing, shuffling groups of dancers gyrating round the parade ground, massed singing and the beat of small drums called madals. I shall never forget the most popular song, or jaunri, which went something like this:

"Eh Jaun, Jaun, pareli anken ma gazeli
Sama Jaunchhu Dehra Dun."

(When you see mascarad eyes winking
You know you are near Dehra Dun).

The future of the Brigade was not decided until more than a year later and by that time I was firmly back in the run of things in the world of the tea bush. Of my fellow company commanders, Peter Richardson transferred to service with British Gurkhas in the 2nd Gurkha Rifles and was subsequently awarded a well deserved DSO for services during the Malayan Emergency. Jim Fillingham stayed on to command a battalion of the 10th Gurkha Rifles during the Confrontation with Indionesia. He was awarded an OBE and, later became largely responsible for the buildup for the Gurkha units magnificent proficiency with small arms. The Brigade has run away with all the major awards at Bisley for a number of years.

I left Wana with a terrible hangover and remember nothing of the convoy to Manzai where I had to spend two nights waiting for the departure of the Heat Stroke Express. I had Bill Thom the bull terrier with me. Generally, intending passengers took one look at him and decided to choose another compartment. This proved a boon throughout the long journey to Ceylon. At Mari Indus where the broad garage started BT accompanied me to the station restaurant for a meal. This was a mistake, for when we returned to our carriage I found a Pathan rifleing through my bags. Fortunately he was seen before he had time to take anything and he disappeared into the darkness persued by BT who shortly afterwards returned with wildly wagging tail and a large smile. After a train change at Lahore and another at Delhi on 6th June, another train took us all the way to Madras through central India via Jhansi, Nagpur, Hyderabad and Kurnool. We stayed the night of 10th June at the Connemara hotel in Madras and during the evening whilst exercising BT on the beach the joy of freedom after so many nights in the train overcame him and he rushed off, disappearing amongst the crowds of fisherfolk and evening strollers. I searched desperately for hours but there was no sign of him. I returned to the hotel for dinner, then resumed the search. As I walked along in the dark something thumped into me from behind. I turned round expecting to have to deal with an attack but, instead, I was leapt upon by an ecstatic, excited Bill Thom, panting and whimpering at finding his master. Next morning we caught the South India Railway's train for Dhanaskodi from Madras' Egmore station. This time we had company, the other occupants of the four berth compartment turned out to be "Bull" Mack, a well known Ceylon planter and rugger player, and a padre. Bull was on his way back to his Dickoya estate after an adventurous war. He produced a bottle of Scotch Whisky - a rare and valuable commodity then - and insisted we three drank it in equal shares. I can't say I enjoyed the twenty hour journey and I am certain the padre did not. When we crossed the Palk Straits next day, and boarded the Ceylon Government Railway's Night Mail for Colombo, I managed to give Bull the slip and BT and I once again had a coupe to ourselves. It seemed strange to waken to the luscious greenery and steamy tropical growth after years spent in the dry and barren wastes of the North West Frontier area. I approached Colombo with mixed feelings.

PART 3

CHAPTER TWELVE

TOWARDS INDEPENDENCE, 1946 - 1948

"Let others sing the praise of wine,
Let others deem its joys divine,
Its fleeting bliss shall ne'er be mine.
Give me a cup of tea!".

From "The cup for me" a song said to have been promoted by
the Victorian temperance lobby.

I arrived back on Spring Valley to find all the planting Staff appointments full. Jock Sutherland was back in charge, and Jack Cranfield had returned to Kottagodde. Colonel Maxwell-Johnstone had left, having handed over my old divisions to Duncan Reith, whilst Arden Oliver was in charge of 3rd and 4th divisions. Reith and Oliver were early examples of Eurasians being given planting charges in company owned estates; a situation which would soon change dramatically. Bob Willett, a new-comer to the island, was on First Division. Bob had been wounded at the battle of El Alamain where he was a sergeant tank commander. His unit was in the forefront of the battle and suffered many casualties. It was the effect of these, rather than his own wound, which led to Bob being invalided Home and out of the army. About a year later, when he was looking for a job in "civvy street", Bob saw an advertisement for a planting vacancy in Ceylon where there was a desperate shortage of suit-able men due to planters away on war service. Being the only applicant, Bob was soon on his way out via The Cape. He had to change ships at Durban and had a long wait to find a berth on a ship heading for Ceylon. However time in Durban was not wasted. The city was noted for its hospital-ity to servicemen. Bob was treated as a wounded hero by the girls there, and made the most of it.

The Spring Valley labour did not know what to make of Bob. He constantly pulled their legs, and at morning muster made all the men move at the double. This was not popular but they forgave him because of his care and kindness to the women and children. Bob kept in touch with his girl friends in South Africa and, eventually,became engaged to one of them. He sent money for her passage to Colombo and arranged a wedding to take place immediately she landed. The story (probably much exaggerated), was that when she came ashore he discovered she was the wrong girl. There was noth-ing for it. All arrangements were made and guests waiting in the church. The wedding went ahead. Actually Dorothy was charming. She was older than Bob and had led a sheltered life in Durban, where the blacks were strictly segregated from the whites. She was quite unnerved by the close proximity of so many Tamils and Sinhalese on Spring Valley.For a long time she would not venture out of the bungalow alone. Dorothy's situation'was not helped by the fact that Bob's only means of transport was his motorcycle and the monsoon weather made pillion riding more than hazardous and uncomfortable. Nor were matters improved by Bob's habit of collecting snakes and other beasties. They turned up everywhere he went, Sometimes round his telephone cable, in a cupboard, or more often in the field when he was amongst his pluckers. News of his hobby soon spread through the estate and the villagers and children arrived by the score bringing their offerings. Bob was also a raconteur of tall stories of his adventures with rabid dogs, wild boar and leopard. Life was seldom dull for us with Bob around, but for Dorothy it must have been awful.

There being no vacant bungalow for me I stayed with Jock and Myrtle, occupying the annexe at Big Bungalow. Bill Thom who accepted the change from Waziristan to Uva with equinamity, suffered the attentions of the resident Aberdeen terriers and Cairns with disdain, but did not take kindly to the pie dogs in the lines. He soon gained their fear and respect. After some skirmishes when he put whole packs of them to flight they decided to give him a wide berth.

I was set to work reorganising the stores and workshops at lst Division. These were in a muddle following the destruction of the main factory by fire in 1943. A new factory was in course of construction 1,500 ft further up the mountain, at May Mallay. I also spent a lot of time supervising work on the construction site, along with the engineers. The erection had been much delayed by shortage of materials needed for the war effort. Both Jock and Jack were long overdue for Home leave. They had been anxiously waiting my return from the services in order to get away. Jock and Myrtle departed for Home in September and I moved into Kottagodde while Jack and May presided at Big Bungalow. The Kottagodde factory was working continuously round the clock to cope with the greater part of the leaf which normally went to the main factory. We found that with careful management we could manufacture more than double the crop which this little factory was designed for.. It meant that the machinery worked twentyfour hours a day, day after day and month after month. In fact the situation continued for over four years. Long hours were spent in the factory and it says much for the staff and labour that no major breakdowns or incidents occured whilst the factory was so overloaded.

Duncan Reith and Arden Oliver were rather upset when they discovered I was senior to them. They regarded me as a newcomer and probably had not expected that I would return after the war ended. They made representations to the Company, to the Ceylon Planters Society, and to the local member of the State Council. However I had remained on the Company's books throughout my war service and my Provident Fund contributions had been regularly paid, so they really had no case.

Writing home on 15th June I told my parents that "the estate looks about the same except there is a huge new factory going up just above May Mallay Bungalow. The country is certainly as beautiful as ever, and if not as healthy it is certainly easier to look at than the barren hills of Waziristan" Something which did impress me was the much healthier appearance of the workers.

Strangely this had been aided by food rationing, whereby bread and flour supplemented the previous monotonous rice diet preferred by the workers. Other factors were the issue of vegetables grown in estate "food production" areas, and the dramatic effect of DDT spraying on the malaria producing mosquito A culicifacies.

Those planters who had been left behind to run the estates, after most of us younger men had departed on war service, found themselves with greatly increased responsibilities. Not only did these older men have the added acreages to look after. Their work was vastly increased by all the necessary war-time measures, often hurriedly introduced, to bring under control practically every item normally used on estates. During this period of stress and strain was added the necessity of providing the machinery for the new Ceylon Estates Employers Federation (CEEF), in all the planting districts. This Trade Union of Estate Owners came into being in order that owners interests could be properly represented in official dealings with the Government's Labour officials and Workers Trade Unions.

During the war years Trade Union activity had greatly increased. Unfortunately the workers unions tended to organise lightning strikes and other subversive activities for political reasons rather than due to any specific grievances against the estates management. A number of large factories were mysteriously destroyed by fire; one of these being Spring Valley big factory. The cause of this fire was never proved but there was strong suspicion of arson. Rumours were numerous. One being that German and Japanese agents were involved. Another said that the engineering firms wanted more work. The latter seems most unlikely as much of the engineers work had been diverted to war production, and materials for rebuilding for civilian purposes were just not available. When I returned in June 1946 the New Factory site had been cut and, thanks to to General Marshall's Lease Lend scheme, materials were beginning to arrive thick and fast. A "sprinkler" fire protection scheme was to be fitted and one of my jobs was to build a water supply tank for this on the hillside above the site.

Whilst the war was still in progress the Soulbury Commission, headed by Lord Soulbury, had arrived in the island to consider constitutional reforms, but it was not until 4th February, 1948 that Ceylon became a self governing member of the British Commonwealth. The Duke of Gloucester then visited the island to lead the ceremonies connected with the transfer of power. It was in the midst of all these tamashas that I boarded the m.v. "Oranje", with my sister Margaret, and sailed from Colombo on Home leave.

But I have run ahead of events. One evening at the end of July 1946 I had an unexpected 'phone call from Kandy. It was my brother Francis whose ship HMS "Alaunia" had put into Trincomalee, homeward bound from the far east. Francis had got a lift and had just two days ashore before "Alaunia's" departure. He was stuck in Kandy. Could I get over to meet him? Those were still the days of strict petrol rationing but fortunately the little Morris Eight tourer left in my charge by Jock Sutherland, had a full tank of petrol and I had a few coupons to spare. So I set off in the early hours and arrived at the Queen's Hotel soon after dawn, to wake Francis and take him down to a huge breakfast. He was looking fit and well and it was good to see him again after his five years of adventurous and distinguished service in nearly all the war theatres. In the days of our extreme youth we had shared many adventures, but I had seen very little of him during the 12 years since I had gone out to Ceylon, After exploring the town and lunching at the Kandy Club I ran Francis out along the valley of the Mahaweli Ganga, to Teldeniya, to give him some idea of the beauty of the area and a glimpse of some tea and rubber estates. It was soon time for his return to Trincomalee and for my return journey over the central range, via the Ramboda and Hakgalla passes, then on to Hangli-ela, Hali-Ella, through Badulla and so to Spring Valley, arriving at my bungalow after 2 a.m. I was up at dawn and ready for a day's work. What it was to be young and fit!

I have already mentioned the increase in Trade Union activity for political purposes. 1946 and 47 were specially trying years in this respect. I will give just one example. The Ceylon Indian

Congress, later known as the Ceylon Workers Congress, together with other unions organised "Hartals", or political strikes, in all the plantation districts, as a protest against the government's acquring Knavesmire Estate, Kegalle, for the settlement of landless Sinhalese villagers. This involved the dismissal and ejection of all the resident Indian workers and their families, who had been employed there ever since the opening of the estate at the turn of the century. The result was a lengthy strike by some 150,000 workers over a matter completely outside the control of the employers. The loss to the industry was great and the affair resulted in much discontent. The European Superintendent was among those dismissed and this was regarded by some as a portent of things to come, despite assurances from the government.

Many of the old planters who had stayed on beyond their time of retirement, due to the war, were now leaving the island. Some of the new recruits were not suited to the job. The old and well tried creeper system had fallen out of favour. Applicants were interviewed by directors in the London offices. Some very good, and some very bad selections were made, but there was not.the same close-knit family influence as before the war. A few who applied had visited estates on leave during the war, and had gained an incorrect impression of a planter's life. They had been entertained at the club, waited on by servants, and were attracted by what they thought would be the life of Riley. It took a little time for the bad ones to be weeded out, but in the meantime they contributed to some serious disputes and a worsening of relations with labour.

On Spring Valley we were embarking on a wide ranging programme of building new housing and modernising selected old type line rooms. Each new cottage dwelling was fitted with its own kitchen and stove, and integral latrine, in addition to the normal verandah and living room. Strangely these improvements were not always accepted by the workers who preferred to live hugger mugger in the old rows. We adopted a process of careful selection of individual families to occupy the new quarters and they gradually came to appreciate the amenities, and so the word was spread. Unfortunately, owing to political interference, and falling profits following punitive taxation, the programme was never completed.

Late in 1945, whilst I was still at Wana, my sister Andrea was posted to India with a detatchment of the Queen Alexandrais' Imperial Military Nursing Service. I had high hopes of meeting her, but after a short stay in Poona her unit was sent off to the far North East Corner of Assam, beyond Dibrugarh on the Brahamaputra river, far beyond my reach. There was still a large, army in Assam and Burma with many of its men requiring hospitalisation from the effects of war and disease. Before I could organise a meeting I had to return to Ceylon and, soon after that, Andrea, and her unit returned to Britain. So we did not meet east of Suez until about nine years later. Brother Louis was already back Home and about to sail for the Gold Coast with a job with the Agricultural Department there. Then in about early 1947, brother Cuthbert (Tupper) and his wife Penny with their two small children Philippa and Bruce, arrived in India. Tupper had a staff job at Deolali, the huge transit camp for British troops outside Bombay. It was his job to see that everything to do with shipping the British troops Home went smoothly. He was far too busy to contemplate a visit to me in Ceylon. Next came news that Margaret, my youngest sister, was on her way out to help with Tupper and Penny's children, and would be with them for about six months. This seemed to offer a great opportunity for Margaret to come on to me, at the end of her Deolali visit. Everything was arranged in due course and Margaret came down by sea from Bombay in the company of Delia. I forget Delias' other name but in Ceylon she met Charles Derry who planted in the remote Moneragala District. They soon married and she became Mrs. Derry.

Margaret's arrival and installation at Kottagodde brought an influx of visits from bachelor planters, I was suprised to find how popular I had become. Tony Perkins would arrive from Serendib on his motor bike. There would be visits from Paul Pern and Ian Mackaye from Demodera. We would be invited over to Mortlake division of Telbedde by Mansergh Hodgson, who kindly loaned his horse to Margaret. She would ride it daily on the estate paths of Kottagodde. The animal always made slow progress on the outward trip, but tried to bolt when homeward bound, usually finishing up on the front lawn with an emergency stop which sent Margaret over its head onto the grass. My cricket and hockey continued but we seemed to stay much longer at the club afterwards. For Christmas 1947 I booked Sinnamuhthuvaran Rest House and invited Dodo Pitts, Charles Edwards, Paul Pern and Ian Mackaye to join us. We did some shooting, mostly snipe, and a lot of swimming and drinking. The Rest House was situated close to the sea on the east coast north of Pottuvil. One evening Charles Edwards and I set out to shoot duck and teal on Rufus Kulum. Margaret came with us to operate the inflatable rubber dinghy which we took on the roof of my little Morris '8'. Mig dropped us off on different islands in the lake and we awaited the evening flight. Soon teal and pintail were coming over thick and fast and quite a few we shot fell into the water. Margaret was told to row out and pick them up, but was shocked to find crocodiles were getting there as fast as she was. Charles Ed and I were in fits of laughter until we suddenly realised that our only escape from the islands depended on the inflatable which, any moment, seemed likely to be punctured by crocodiles teeth.

Ian Mackenzie was back, having survived as a prisoner of war to the Japanese Imperial Army. His unit, a battalion of the Jats, had entered Siam briefly, in fulfilment of their assigned role. Then they had retreated the whole length of the Malay Peninsular to Singapore island, fighting all the way. He had been put to work on the "death railway" and suffered privations though he would never speak of those times. Mig and I called on him and his new wife, Peggy, at Rangbodde Estate off the Ramboda Pass. Ian was building an Enterprise dinghy using local woods. Bob Carswell was also back and had recently taken charge of Newburgh Estate, Ella, from Robbie Gray who had moved to Attempettia. When serving in Assam, Bob had met Marian Robertson, a VAD serving at Tansukia and later Rangoon. They married and, later on, after I too had married, we enjoyed many childrens parties at Newburgh

At Kottagodde I was enjoying my first independent charge of an estate and tea factory. With the factory working flat out round the clock I was fortunate to inherit Jack Cranfield's excellent organisation.. I also had old K.Sinniah, the teamaker, who had not a word of English. He was one of the old type of staff who knew his job backwards and was utterly reliable, and knew how to get the best out of the labour. It was a joy to visit the factory in the cool, early hours, of the morning to see the loft workers "knocking down" the withered leaf; to savour its cyder-apple like scent, and to check-weigh the leaf ready to chute down to the rolling room below. There it was twisted and curled in the rotating rollers in the cool humidified chamber. With the squeezing out of the juices the scents grew stronger. Eventually the fermented leaf was fed to the driers. All the machinery was driven by the huge single cylinder Tangye engines, with monstrous fly-wheels speeding round: the piston shooting in and out of its casing. The steady chuff-chuff of the exhausts could be heard plainly a mile or so away. I always enjoyed watching the engine drivers, Murugan or Carupiah, pouring oil onto the moving parts from their longspouted oil cans. The greaser would be slapping belt paste onto the whirring, flapping belting which drove the main shaft. This shaft extended the full length of the building. It had numerous pulley wheels positioned along its length. From these many smaller belts extended upwards or downwards, to drive the seven rollers, three roll-breakers, the driers and all the various machinery in the sifting room. They also moved the two great central fans, which sucked in the outside air, mixed it in the bulking chamber, with hot dry air from the driers exhausts, and pushed the carefully controlled mixture over the green leaf which was thinly spread on the tiers of "tats" in the lofts. There was a special atmosphere in the old belt driven machinery which was missing in later days when electric motors were installed, and each machine became controlled by the flick of a switch.

At Kottagodde,in those days, the main shaft had to rotate at exactly 123 revolutions per minute, to achieve the correct draught and temperature in the driers, and the right speed of the rollers, fans and sifters. The engine speed had to be just so. A loose belt would slip on the pulley wheels and cause the main shaft to loose speed, and a broken main drive belt brought everything to a halt. This meant the drier furnaces had to be raked out to avoid severe overheating. All these things had to be watched for twenty four hours a day, seven days a week, throughout the four and a half years it took to build the new factory. Thanks to the devotion of Sinniah and his workers it was seldom necessary to call in the engineers from Badulla.

Packing the made-tea for export took place every few days. The nailing coolies assembled the ply-wood chests and fitted the aluminium foil linings, after this they would be stacked in the drying room to remove all semblance of moisture. Samples of tea would then be tested for moisture content, to see whether a "final fire' was necessary. The primary object of firing the tea is the arrestment of fermentation. However after it is fired the tea continues to absorb moisture from the air during the process of grading, and again during storage and transportation. To avoid complex explanations it is enough to state that the safe limit for the moisture content of fired tea is in the region of 3%. This having been achieved the storage bins would be opened and the tea piled in a pyramid-like heap on a bulking cloth in the centre of the room. Then bulking with wooden shovels would ensure that the tea from each day's make would be thoroughly and evenly mixed together. Any hint of the fluctuations in leaf colour, grading size and liquor quality, which inevitably occurred when the weather differed during day to day harvesting,and manufacture, must be prevented.

After each chest was loaded on a vibrating packer, to exactly the right weight, the packing labourers would sieze stencilling equipment and enter the tare, net and gross weights of each chest. The name of the estate, the number of the invoice, serial number of each chest, the grade of tea and the port of destination would also be marked.. The chests would next be check weighed and counted onto lorries and despatched to railhead at Badulla. Samples taken from every chest of the invoice would be laid out for tasting, examination and comparison with recently received samples from neighbouring estates. Samples would also be posted to Colombo and London for valuation and report by the tea brokers. Careful attention was given to the reports for any criticism which might indicate need for changes in handling and manufacture. Our aim was to keep ahead of our neighbours in quantity and quality. There was never a let-up, and no room for complacency. When the new factory came into operation a good deal of the work was automated.

I had occasional news from the Fowlers who had left Yapame,on retirement in the spring of 1946. Flabby died of a heart attack in a train in 1947. His death left a great sense of loss. 'Saibo' Sim was still on Kehelwatte, but would soon be moving to the less demanding charge of the Passekudah coconut estate on the east coast. He first tried a spell at Home. He said he felt like a stranger in his family home in Scotland, and the winters were horrible. 'Have-a-dram' Ruxton was still at Hopton; Desmond Pelly was on Passara Group, 'Gus' Hall administered Dammaria, Aubrey Clarke managed El Teb. I used to call there after tennis at Passara, as did several other admirers of his tall blonde.daughter whose name I forget. She married an officer in a Gurkha regiment and both her brothers took up planting. George Kent-Deaker was in charge of Gonakelle estate. One of his new SDs, after the war, was Charles Edwards, just demobbed from the Royal Navy. Charles was ordered by Georgie Kent-Deaker, to attend morning muster daily. This seemed a bit much to Charles who was no teen-age creeper straight from school but a seasoned naval officer. So he devised a method of avoiding muster whilst making GKD think his orders were being obeyed. Careful enquiry and observation had convinced Charles that GKD listened for the motor cycle which Charles rode to muster. On hearing the noise he was satisfied and went to his office behind his bungalow. So

Charles gave his Appu (cook) lessons at riding the bike, then having trained him well set him to ride round and round the lawn while Charles relaxed in bed for another hour. The Macdonalds were still on Cannaverella. Mrs MacD was reputed to be the power behind the throne. She designed the new bungalows and ordered the factory work. The Carey family still owned Pingarawa and their relation, Wilfred Rettie,was still the VA at Spring Valley and Glen Alpin. Nigel Bannerman was on Ledgerwaite with a new SD, John Davis, and the Wills were at Glen Alpin with Tommy Williams and Pat Fincher competing for the position of heir to the property. Both Percy Will and Nigel Bannerman became Managing Directors of the. Spring Valley Company during the 1950's, and their mantle was taken on by Tommy Williams in the 1960's.

During a hockey match in Badulla, shortly after Mig's arrival, I was bumped into by the bulk of Gerald Du pre Moore, resulting in a broken and dislocated left collar bone. It was badly set at Badulla hospital and Jock advised me to go for treatment at the Joseph Fraser Nursing Home in Colombo. Kind friends agreed to put Mig up during my absence, which was for a few days only, as I discharged myself early, thinking that such a minor injury scarcely demanded hospitalisation. Back on the estate it was time to get ready for Home Leave. Jack was due back any moment and would want to move into Kottagodde. I got cabins for each of us on the Nederland Lloyd's m.v 'Oranje', sailing between Java and Holland, and calling at Colombo, Port Said and Southampton. Colombo hotels were pretty full for the Independence celebrations and the Duke of Gloucester's entourage, but we managed to get rooms at Mount Lavinia and, after a few hectically social days, and last minute applications for a permit to export an elephant's foot waste-paper basket, we set sail for Home. Mig had accquired the foot from some planter friend and insisted on taking it loose in her cabin. On arrival at Waterloo it found a place on top of a luggage trolley, causing much comment from the porters who, in those days, were still numerous at the London terminii. This was my first Home Leave as a civilian and I was looking forward immensly to a six months holiday. In the event it lasted a whole year and had historic consequences for me. The past two years had been difficult ones. The future was uncertain, but I still loved Ceylon with its unique charm and atmosphere, but something seemed to be missing.

CHAPTER THIRTEEN

HOME ON LEAVE - 1948

"Come live with me, and be my love, And we will some new pleasures prove Of golden sands, and crystal brooks, with silken lines, and silver hooks"

John Donne.

The M.V. "Oranje" docked at Southampton early on third March, 1948, and I arrived Home, with Margaret, the same day, to the ever warm and loving welcome of my mother and father. During the following weeks I went to London to meet the directors of the Spring Valley company, now back in their old London office in Thames House, Queen Street Place, after spending the war years evacuated at Woodmansterne in Surrey. The genial J.W. Scott, late of Glen Alpin, was the Managing Director, and the famed Andrew Young, was Chairman. I was taken to lunch at Sweetings Restaurant and rather reluctantly offered a glass of ale. These careful Scots always made you feel this was a great extravagance. The rebuilding of the war damaged city was going on at a great pace, and it would not be long before the company transferred its headquarters to a new block of offices at 21 Mincing Lane.

My brother Cuthbert (Tup), had a posting to the War Office and we arranged to meet at the Hyde Park Hotel, where he was billeted. Francis joined us there. He had recently been demobbed and was swatting for his Masters Ticket at a navigation school. He introduced me to some of his London haunts in Clubs and Pubs, where he seemed to be well known and popular. He travelled on his bicycle whilst I went by bus or taxi, or walked. Whatever the means of my transport Francis always arrived first. Tup soon moved, with Joy and their young family, to a house in Chingford in the Epping Forest. I stayed with them a couple of times, enjoying walks through the forest or taking the children on boats on a lake. Margaret was in the process of finding a housekeeper's job at a London Hotel, and Andrea was at the Bristol Royal Infirmary, having recently come Home from nursing with the Q.As in the Suez Canal zone. John was Executive officer in HMS "Conway", in the Menai Straits, off Bangor, and Guy was serving his last term in the ship as a cadet.

In April I hired a small car and drove to Bristol to visit Andrea. Petrol was still severely rationed but, as an overseas visitor, I was allowed a special quota. The drive to Bristol was easy as the roads were almost empty. Having arrived there, however, I couldn't find any accommodation. Bristol had suffered badly in the Blitz and hotel beds were at a premium. Eventually I tried the YMCA. This was also full up. The YMCA Secretary suggested I tried the police station where, to my surprise, I was given a clean cell with coarse,but clean bedding. A kindly policeman brought me a cup of hot strong tea on each morning of my two night stay. Andrea was well but very busy, so I arranged to visit my favourite Aunt at Nightingale Farm, in the village of Brent Knoll, Somerset.

170

Aunt Maud made me very welcome in her delightful home. She missed Michael, her eldest son, who had recently migrated to British Columbia. June, her daughter, who I had known when we were small children, came down from London for the week-end. I think she was at an art school. Raymond, the youngest son lived at Brean, a few miles away on the coast, where he was running a small horticultural holding in partnership with Michael Holmes, a cousin of my future wife. They specialised in daffodils and tomatoes.

Aunt Maud's husband, John Perks, had, died some years previously. Thr Perks family owned an agricultural tool-making business at Wolverhampton which manufactured Chillington brand tools for the plantations of India, Ceylon, Malaya and other tropical areas. I remember, as a small boy, being shown photographs of his visits to Ceylon, including pictures of the ''buried cities'', little knowing that I would later spend half a life-time in the island. I still have two Chillington brand mamoties.

Brent Knoll was an attractive village in those days, with its long straight main street running along the flat land below the Knoll. A side road ran off to the north, onto the lower slopes of the hill where stood (and still stands), the ancient St. Michael's church, the Vicarage, The Manor, Ballcopse Hall and a couple of old farm houses. Behind the church a footpath leads steeply up to the top of the knoll on whose summit is an ancient earthwork encampment. From the top there are fine views towards the cities of Wells and Glastonbury, Bridgewater, Weston-Super-Mare, the hills of Exmoor. Sometimes the coast of South Wales is visible.

Aunt Maud announced that in the house opposite lived a retired naval commander, his wife and pretty daughter. She intended inviting them over for a drink if the daughter, who was training for her SRN, in London, came home for the week-end.

Staying in the village, at that time, were Donald and Carey Marley, on home leave from St. James' Estate, Haliela, about six miles as the crow flies, from Spring Valley, but about 40 minutes away by the winding mountain road. Donald had been in my section of the CPRC when we were mobilised to Trincomalee at the beginning of the war, so I knew him well. Carey was a very pretty girl, and they had two charming children, Philip and Susan. In Brent Knoll they were lodged at Croft House, at the far end of the village, with Carey's parents Frank and Lilla Jepson, also late of Ceylon. Frank had held the,important post of Director of the Royal Botanical Gardens at Peradeniya outside Kandy. He was known as a bit of a ladies man.

My 1948 diary has no entries until Sunday 18th April, when the following occurs :- "Met Rosemary Prescott-Roberts at Marleys in Brent Knoll"; then on the 19th the entry reads:- "Cinema in Weston-s-mare - 'An Ideal Husband'. Sat next to Rosemary". These laconic entries fail to reveal that I was walking past Croft House, on my way to the Fox And Goose, for a drink that Sunday evening, when a voice called over the garden wall, "come in and join us, Charles". A game of tenni-quoits was in progress on the lawn and I noticed one of the players was a very pretty girl, Donald and Carey introduced me to the old couple and I sat with them, watching the graceful girl on the court. Then I was introduced to Miss Rosemary Prescott-Roberts and partnered her in the next game. As soon as I looked at Rosemary I was completely bowled over. She was exactly like the girl I had sometimes met in my dreams. We immediately got on famously and I was determined to see more of her. It turned out that she was the girl mentioned by Aunt Maud, who lived at Thorncote, opposite Nightingale Farm. Some of the electricity between us must have been communicated to

Carey, because she invited us both to join her family in a bus trip to Western-Super-Mare next day, to see the film of Oscar Wilde's "An Ideal Husband". The bus ride and the film proved a great success and I determined to meet Miss Prescott-Roberts again as soon as possible.

The following morning I was due to go to Okehampton, to stay with Carl and Minnie Drieisma (see chapter 1). I found I had to wait an hour on Exeter station, so I telephoned to Thorncote. Mrs P-R answered the call and kindly gave me her daughter's London telephone number at St. Clement's Hospital, Bow Road, in the East End, where she was training for her State Registered nursing qualification. Telephone calls were costly in those days and, outside the metropolis, direct dialling connections were rare; nor were telephones then readily available to staff in hospitals and business premises, so most communication was carried on by letter. The post was very fast and reliable, a letter costing just 2.5d (old pence). There was no second class post.

On 23rd April I wrote as follows:

"Dear Miss Prescott-Roberts,
I hope you will forgive me for telephoning from Exeter the other day, and for asking your mother for your London telephone number.

I had hoped to be giving a small party, to celebrate my leave, in London on 29th April, and did not know enough girls to make up the party. Now however the party will have to be later on, as two Ceylon friends cannot make that date.

However I shall be in London on the 29th on my way to stay a couple of nights with friends in Surrey and. if you would care to, I would like you to do a show either on that day or on May 3rd. I think I can get either my brother, or my sister and another friend to come along too, but might have to call on you for another girl at the last minute.

If you can come on either of those dates I hope you will let me know by return, and suggest a time for me to phone and a place to meet.
Yours sincerely,
Charles F. Brooke-Smith"

Waiting for a reply seemed to last for ever, but on the 26th a pale blue envelope with large clear writing was delivered to me at Brook Cottage by our faithful postman Mr.Bloom Buckett; It read:

Tel; ADVANCE 3983.

St Clement's Hospital
2A. Bow Road, E3
25th April 1948

"Dear Mr.Brooke-Smith,

I returned from leave this evening and found your letter waiting for me, for which many thanks.
I am sorry I was out when you phoned. I was busy selling flags for the Children's Appeal.
It is very kind of you to invite me to a show and I would love to come. I could actually manage either the 29th or the 3rd, but 29th would be better as there is more chance of my getting off punctually. Would 6pm be too late? (6.30 in town). The best time to phone is about 12.30pm or

before 8.30am! But I think that is asking rather a lot. I could easily provide another girl if necessary.

Yours sincerely,

Rosemary Prescott-Roberts".

I managed to get good seats for "Bless The Bride" at the Adelphi, on 3rd May and called Miss P R before 8.30am to arrange an RV. I got brother Francis to book a table at Manettis' Restaurant, for after the show, and he came to partner the other girl, Margaret Sandover (Sandy), who was also doing her training'at St. Clement's.

The play, the music and songs, were just right for the occasion, and the evening was a huge success. There was a slight difficulty when it came time to leave Manettis, as taxi drivers were reluctant to go all the way to the East End, and it was late for buses and the tube. The Problem was solved by offering the taxi driver a swig from my whisky flask (an old dettol bottle). Scotch whisky was then desperately short so the inducement was eagerly accepted, resulting in the girls being rapidly conveyed to the hospital where they arrived just in time to avoid recrimination.

Such were the civilised preliminaries to courting in those far-off days.

The following day I rang the hospital and invited Rosemary to come to "Annie Get Your Gun", a Rodgers and Hammerstein musical then all the rage in London. There were no seats left in the body of the theatre (The London Coliseum), but I managed to get a box. I asked Desmond Pelly, who was on leave from Passara Estate,to join us and we met Rosemary in the foyer. She brought along Fiona Baker, an old school friend. After the show we walked along to Hatchett's restaurant for supper, and I took the opportunity to ask Rosemary to join me in a trip to Hampton Court the next day, Whitsunday, 16th May. We first went to Richmond then on by river steamer to Hampton Court. We got back to central London in time for a concert at the Albert Hall. Basil Cameron conducted the London Philharmonic in a concert of music by Tchaikovsky. The next Sunday I hired a little Morris '8' for a drive out to Ongar and Bishop's Stortford. We bought strawberries and ate them in a field. My diary for that day (23rd May), reads; "Proposed near Ongar- Some Hope, Saw "The Great Waltz" in the evening".

Although my proposal had been turned down - but very gently - there was no let-up in my pursuit. A week later I took Rosemary to see another Rodgers & Hammerstein smash hit musical, "Oklahoma', at the Drury Lane theatre. This time we were on our own and had supper afterwards at the Cafe de Paris. We continued going to plays, concerts and cinemas. Sometimes we had supper in the Buttery at the Royal Overseas League, where I was a member. After one such supper on a very wet evening I delivered Rosemary home to St-Clement's by taxi, returning to the West End in the taxi to search for a bed for the night. There had been no time earlier and the Overseas League was full, as were my other usual haunts. I walked the streets into the early hours in that pouring rain. Walking in London on a wet night is far worse than pushing through sopping jungle in Ceylon. Eventually I returned to the Overseas where the night porter took pity on me, ushering me into the vast Hall of India, giving me a blanket, and promising to bring me tea and hot shaving water at 7am. I slept soundly for the remaining part of the night on a huge sofa and shaved in front of an immense gilt framed mirror. Another night was spent at the Great Eastern Hotel where I shared a room with a fat Yorkshire commercial traveller. All this made me try to find more regular lodgings closer to Rosemary.

During June Rosemary, Sandy and Pauline Manwell transferred from St.Clement's to St.Alfege's Hospital, Greenwich, for the latter part of their year's training. Having been senior VADs in the services during the war they had considerable practical nursing experience, and were now undertaking a shortened course to qualify as civilian State Registered Nurses. In those days nursing staff "lived-in" at the hospitals. They worked studied, messed and slept there, and the rules demanded that on "days-off" and "half-days", they should be back by 11pm; though late passes were sometimes issued. This system may have had its disadvantages, but at least nurses were not distracted by having to do their own housekeeping and cooking. They were each provided with a room and basic furniture. My attentions to Rosemary must have distracted her from her work and studies but she always had interesting, and often amusing, stories to tell about her patients who were evidently devoted to her..

I booked into a place called The Khandala Hotel in Montpelier Row, overlooking the Blackheath. At the same time I was searching for a temporary job to earn some money as my resources were dwindling rapidly. I called at a variety of organisations, such as motor-car manufacturers, with a view to driving new cars from factory to ports for export. Car transporters were yet to be introduced. I tried for Commissionaires posts, and to be a stevedore in the docks (to break strikes), etcetera. I soon discovered that a Trade Union membership card was "sine qua non" for any of them. As for strike breaking, the employers were much too scared of the unions even to consider looking at a potential 'blackleg'. That was a time of serious dock strikes with the strikers holding the country to ransom and making the already short rations even more scarce. In the end I got the Spring Valley Company's Secretary, McKerron Young,to arrange a course (unpaid), for me with Wilson Smithett's the tea brokers. At least it kept me occupied for a few days and I had a glimpse of what went on behind the doors of the London Tea Market.

The Khandala, and Montpelier Row, had seen better days. These classical buildings showed signs of bomb damage, and many of the old people who comprised most of the lodgers were probably refugees from bombed out homes, or had retired on pre-war pensions which now failed to support the standards to which they were accustomed. Many were pathetically eager to talk to a newcomer, especially a young one for a change. It was run by a couple with the Welsh name of Jones though I am sure they were Irish. In a letter to Rosemary I told her "as a matter of fact this dump could be a lot worse. At least the inmates are genteel if faded, and it is CHEAP. My bed is certainly bumpy and about eighteen inches short".

Blackheath and the beautiful Greenwich Park were on the doorstep, and a walk to St.Alfege's Hospital took only ten minutes. The summer weather and long evenings made strolling a pleasure. What could be better than having a charming companion to share the view of the river with Wren's classic contribution of The Royal Naval College, the Hospital, and the Maritime Museum, from a path or seat in the lovely flower gardens of the park. But there was a fly in the ointment. My six months furlough was rushing by. I was due back on the estate by the end of August which would mean sailing away early in that month and we were already into July. Rosemary was determined to complete her training and pass her examinations before making any decision.about the future, and she appeared to be a most determined young lady. There was only one thing to do. i.e. make a bold application for a three months extension of leave. Surprisingly this was readily granted. I had been abroad for over twelve years before the company was obliged to pay any passage money or leave pay, so it was not altogether unearned. Anyway I was thrilled and very grateful to the Board of Directors.

At the end of June I had moved from the Khandala to Gordon House in Blackheath Park Road. This establishment was run by The Bank of England. How I came to be allowed to use its facilities I remember not. I do recollect that I had a spacious bedroom and the shared use of a servant. There was a bar and a well furnished lounge, a large garden, and playing fields across the road. The house seemed to be little used except perhaps at week-ends when I tended to be away. I even arranged for my youngest brother Guy to stay for a night before he set off on his first voyage as an apprentice with Ellerman's City Line.

One Sunday I met Rosemary outside All Saints Church, Blackheath, for a matins service. I took along a beautiful red rose which I noted hanging over a garden wall on the way. After the service I was invited to visit her home and parents at Brent Knoll, for a week-end in August. Of course I hired the small car and we drove down. It was a wonderful drive on almost deserted roads. I was taking full advantage of the extra petrol allowance for overseas visitors. The P-Rs were marvellous hosts. Rosemary took me to visit various friends in the village and we climbed The Knoll. Andrea came down from Bristol for a day. She made a great hit with Rosemary's father, and Aunt Maud asked us to a party she gave for June, who became engaged to Carey Marley's younger brother Clifford Jepson.

Life during the spring and early summer that year seemed like a novel by Dornford Yates. The sun was always shining and the countryside looked perfect. Everyone smiled and I had the most delightful girl in the world for company.

We drove back to London taking a picnic which we enjoyed in a field full of wild flowers near Hatesbury. Before we parted Rosemary announced that if she were to get her State Registration she would have to work very hard. Her studies and practical work would allow her no free time, so she would be unable to see anything of me over the coming weeks. I caught a train to Woodbridge. It was Sunday and the train seemed to take for ever, stopping at every station along the line, and in between.During this depressing journey everything seemed to change. I began to wonder. My happy optimistic mood changed to one of deep depression. How could I ask a girl like Rosemary to bury herself in the wilds of Ceylon, miles from anywhere, among alien people of a strange culture, with no female companionship; in a world dominated by men, at risk from strange tropical diseases, and with no social life or cultural activities such as are common in Europe. My first spell in the East had lasted ten years before I got Home. Although it was unlikely to be repeated, it was always a possibility. Anyway the current arrangement for leave once in five years was far too long. It would be the end of 1953 before we would get Home again. So much could happen during that time. We could even turn up with a family of five.! I arrived home in a depressed state. My bad temper and silences must have puzzled and pained my parents who knew nothing about my pursuit of Rosemary.

Being too restless to spend much time at home I wangled an invitation to stay a long week-end with my Aunt Joan and her husband Adrian Lesser at their lovely home Rick Barton, near Chideock in Dorsetshire. I went on from there to visit another Aunt at Rowney House, Wimbourne. Aunt Beatrice, known to all the family as Aunt Mouse, was a great favourite. She had married, late in life, the Liberal member of Parliament Athol Randall who, amongst various other accomplishments, had played croquet for England and was a fanatic of the game. Needless to say I was initiated into the mysteries on their well kept lawns. They also sported a real live butler and a valet whose disapproval was obvious when he unpacked my shabby suitcase. My next expedition was to Little Stretton, a tiny village in Shropshire, then quite unspoilt, lying at the foot of the Long Mynd in the Shropshire Hills. My brother Cuthbert had recently bought Mynd House. His wife Joy and the three children

Miss Rosemary Prescott-Roberts, 1948.

6th November 1948.

Margaret Sandover
(Bridesmaid).

The Groom and Bride.

Lt. Col. C. Brooke-Smith
(Best man)

were settling in. I was kept busy with painting and repair jobs, picking the huge crop of raspberries, and helping to look after the infants. After a few days I moved on to Beaumaris in Angelsey. My brother John was executive officer aboard H.M.S. "Conway", the cadet training ship moored in the Menai Straits, off Bangor. From my lodgings in Baumaris I paid several visits to that wonderful old wooden three decker ship of the line, including attendance at Sunday Divisions followed by cocktails with Captain and Mrs. Hewitt in their spacious stern quarters. John arranged for me to be one of the crew aboard a Mersey Merl class thirty foot yacht during the local regatta. The owner skipper was a fine old boy who was commodore of, I think, the Royal Dee Yacht Club. After races we would meet on the terrace of the Bulkley Arms Hotel, with its glorious view across the Menai Straits to Snowdonia. In the evenings there would be lively parties in the local pubs. My rather dubious speciality by way of entertainment was to drink a pint of beer whilst standing on my head.

The next destination on this Round England Tour was Dinnington in South Yorkshire. My mother's youngest brother, Haffey Bean, was rector. After a couple of nights in the comfortable rectory, getting to know my teen age cousins Nigel and Thea, I headed back to Suffolk. All the journeys were by train. The railways still reached into almost every corner of the land. They had just been nationalized, on lst. January 1948, and as yet showed almost no signs of the changes which were to come under state ownership.

Rosemary's letters followed me to all my destinations and I wrote to her daily. One of my letters contained an "ultimatum" saying that there was little time left to us; a decision, one way or another, must be made very soon, because when I left for Ceylon it would be like passing out of her life. The alternative was for me to resign my steady job out there and look for work at Home. I asked her to come on a visit to Brook Cottage to talk things over and meet my parents. To my great Joy my invitation was accepted, with a proviso that "you are not even to think of giving up your job in Ceylon. That's my ultimatum". So on Tuesday 24th August I met Rosemary at Ipswich station and we went to the Mikado Tea Rooms in Tavern Street for a calming cup of tea before the ordeal of arriving at Brook Cottage. On the way I stopped the car in a lane near the village of Great Bealings, and there Rosemary agreed she would become another Mrs. Brooke-Smith. Thus I was able to introduce her to my parents as my fiancee, even though by only a few minutes. John and Francis who had been out sailing arrived soon afterwards and seemed sufficiently pleased to mob me. Mum and Dad were delighted and immediately accepted Rosemary as one of the family. It was quite an ordeal which Rosemary took in her stride. Even the family prayers in the evening and the grace before meals were accepted almost without comment. We were left to ourselves a good deal, so were able to walk in the lanes and talk over our plans for the future. Rosemary wanted to be married in England, at Brent Knoll, and to travel out to Ceylon with me. She wanted the wedding to be after the publication of the SRN examination results so that she would have a mind free to concentrate on the wedding arrangements and the preparations for the voyage. This meant that I would have to apply for yet another extension of leave. However I was confident that if I could get the directors to meet Rosemary they would be favourably impressed and my plea would most likely be granted. This was duly arranged and after a brief visit to the office J.W.Scott, (Uncle Scottie), the Managing Director, took us out to lunch at Sweetings Restaurant. This was a rare privilege and we were even given a drink. However it took some time before we heard that the additional leave was granted - but without pay - though first class passages were reserved for us aboard the Bibby Line's M.V. "Warwickshire" due to sail from Liverpool towards the end of December, subject to cargo loading arrangements. The cost for the 24 day voyage was £90 each.

Rosemary carried on at St.Alfege's until the end of September. During off-duty hours I took Rosemary to the Army and Navy Stores in Victoria to choose an engagement ring. Good quality new jewelery was not available because of the war, but we found a very nice gold ring with a single blue sapphire surrounded by small diamonds. The only wedding rings available then were nine carat curtain-ring like objects. However it is lasting well all these years later. Extra clothing coupons were applied for towards the wedding trousseau and several friends donated some from their own limited rations. The hospital matron was not pleased to receive Rosemary's notice but on hearing that she would be going to live in Ceylon dismissed her with the comment "it's a ferrtile country nurse".

We paid another visit to my parents at Brook Cottage and managed some autumn sailing in "Cape Pigeon", before she was laid up from her mooring in the river Deben off Everson's boatyard. There was a drinks party in our honour, attended by family and friends, and most of Hasketon. The wedding was arranged for 6th. November at St. Michael's Church, Brent Knoll. It was quite difficult to limit the guests to about seventy and an anxiety for the P-Rs who hoped somehow to fit everyone into Thorncote for the reception. During my stay of nearly a month, prior to the wedding, I had some amusing yarns with my future father-in-law. It emerged that he had been a midshipman aboard the battleship HMS "Revenge" at the battle of Jutland in 1916. "Revenge's" Commander had been H.H.W. Hughes, a great friend of my father's and a favourite of our family. He was known to us as "Umpo", a name acquired during his service on the East Africa station during the first years of this century, when he had impressed some local chiefs with his great strength and brilliant shooting. Both Hughes and Commander P-R were unfortunate to fall under the Geddes Axe in the early 1930's when our naval strength was drastically cut in the name of economy and political expediency. In so far as petrol rationing allowed we used the 1934 Hillman Minx to visit Bridgewater and Weston Super Mare, but most of our local journeys were by bus or Shank's pony. My memory of those days is fogged and, since we were together, there are no letters to help.

At the wedding in Brent Knoll's historic parish church Canon Wingfield Heale, the vicar and lately chaplain to the Queen, was assisted by Bill Richardson shortly to become Chaplain of The Fleet. He and his wife Joss were very old family friends. Somehow we got through the reception and away to Brent Knoll station. In those post war years of shortages the,system for seat reservations had been abandoned. You just had to take pot luck in the inevitably crowded trains. We put our luggage in the guards van and walked the corridor until we found two vacant seats. I put my small bag up onto the luggage rack and, to our great embarrassment, a shower of confetti descended onto the heads of the passengers below. At Bristol a kindly ticket collector found an empty carriage for us and locked us in so that nobody could disturb us. At least a month before the wedding I had asked Cmdr. P-R to reccommend a good London Hotel for the start of our honeymoon. When no reply was received I began to get anxlous, then a few days before the wedding my letter was returned marked "This hotel closed down at least twenty years ago". Fortunately I was able to make a last minute reservation at The Grosvenor, Victoria, then a very comfortable establishment. It was also convenient for our next destination, The Burlington at Folkestone, from where Rosemary was able to show me some of her childhood haunts at nearby Hythe, and we could easily visit the twin Aunts Eve and Hilda Bean (of QAIMNS fame), at Deal. We also saw brother Louis and his family who had just arrived at Margate on leave from the Gold Coast.

Back at Thorncote we were kept busy packing our belongings and the masses of wedding presents. These were despatched to Liverpool as cargo for the "Warwickshire". Christmas was fun with carols round the village. The choir performed from a horse drawn wagon lit by hurricane lanterns. The next few days were occupied with farewells to relations and friends, then, at last we were on our way. The train was an hour late arriving at Liverpool but we were cheered by the warm and comfortable Exchange Hotel where we spent a night, and most of the morning of 29th December before boarding "Warwickshire in the evening. It was cold and wet and windy and there was a lot of bustle and confusion. An old sailor patient of Rosemary's had somehow got wind of our departure and came aboard to wish us well. We were shown to a comfortable and roomy cabin on the upper deck and found it full of flowers and 'bon voyage' telegrams. The weather was bad from the very beginning. My diary says "Strong gale. very rough. Trunks sliding all over cabin. No sleep" Fortunately neither of us suffered from "mal de mer" but the violent movement quite frightened us. When we entered the dining saloon for breakfast we found it half awash, with sea water sloshing across from port to starboard and back again each time the ship rolled. The skittles on the tables were insufficient to hold the plates in place so we tried to eat from our laps as best we could. Stewards were swept off their feet by the water when they tried to cross the saloon from the galley. Fortunately perhaps,we were almost the only passengers there.

Later that day we lay hove-to in the Bay of Biscay as huge fifty foot high waves bore down upon us. The lounge was virtually uninhabitable with the furniture charging about with each roll of the ship. We heard that the world's largest ship*, and nearly all other shipping had delayed their departures from British ports. Festivities to see in the New Year were cancelled, not that anyone was likely to feel like celebrating. The dawn of lst. January 1949 saw us surrounded by stormy seas but the sky seemed to be clearing and we were steaming slowly ahead towards the south. It was not a propitious start for Rosemary's new life. We had more storms in the Meditarranean, and were poisoned by prawns after navigating the Suez Canal. We had had a brief trip ashore In Suez, with a visit to Simon Artz Emporium, and were amused by the bum-boatmen and their patter about Messrs Mc. Gregor and Mackintosh.

The British Colony of Aden was the second and last port of call before Colombo. Its barren brown rocks came into sight at dawn and as soon as the gangway was down I took Rosemary ashore to Aden's only hotel, the Crescent, where I got her scrambled egg sandwiches, a delicacy she had never before tried. The average annual rainfall was said to be only two and a half inches. The brown rocks, dust and heat tended to confirm this, although it was the coolest time of year. Actually Aden was a popular station for service people and other expatriates. We went out to Crater where, much to Rosemary's delight, a small Arab girl presented her with a bunch of purple flowers. Her grandfather had called at Aden in 1870 on his way to join the Indian Medical Service, and had written "We did not think much of Aden and I don't suppose anybody else ever did". He mentioned the Arab boys who swarmed round the ship diving for coins thrown into the sea by passengers. They were still at it in 1949. Nothing seemed to have changed. The Crescent Hotel was exactly as I remembered it in 1935, even down to the scrambled egg sandwiches. We did not then forsee just how great would be the changes during the next twenty years.

Another storm blew up in the Indian Ocean. It arrived out of the blue and there was no time for orders to close the lower deck's portholes. We were walking along a companionway, to the dining saloon, when a wave came through an open cabin door with, floating on top of it, a baby in its cot.

It was not all bad weather. There were great moments; standing at the rails watching the sunsets; looking for the "Green Flash"; observing schools of porpoises leaping and arching in unison across our bows, and far outstripping our ship as they dashed ahead and over the horison. We saw flying

*The "Queen Mary".

fish fluttering out from the ships' wake, and we played deck skittles with teams made up from Colombo versus Rangoon passengers, and against the crew. We were practically penniless, I being without pay and having had an expensive leave. We couldn't even afford to buy the duty free drink.

Our twenty four days of isolation from the rest of the world ended when the ship dropped anchor in Colombo . We boarded a harbour launch which landed us alongside the old passenger jetty. It was there that I handed Rosemary ashore onto the soil of Ceylon. Our Home was on Spring Valley until the New Year of 1971. There were good and bad times, but for us they were 22 happy years. Perhaps one day our children will tell of it.

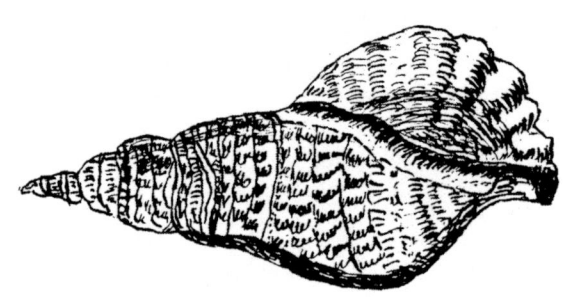

BIBLIOGRAPHY

ANNUAL OF THE EAST	1932 - 1933
BASSETT R.H.	Romantic Ceylon (1929)
CAVE H. W.	Golden Tips (1900)
CHENEVIX-TRENCH C.	Viceroys Agent (1987)
C.H. & D.H.W.	Udday Uppar (1920)
DARNTON IRIS	Jungle Journeys in Ceylon (1975)
FARWELL BYRON	The Gurkhas (1984)
FERGUSONS CEYLON DIRECTORY	(1969 - 1970)
FORREST P. M.	A Hundred years of Ceylon Tea (1967)
GREEN L. B.	The Planters book of Caste & Custom (1925)
HUXLEY GERVAS	Talking Tea (1956)
KEBLE W. T.	Ceylon Beaten Track (1940)
KEEGEL E.L.	Monograph on Tea Production in Ceylon (No. 4) (1958)
MACMILLAN H.F.	Trophical Gardening and Planting (1948)
MAINS LT. COL. A.A.	Field Security - Very Ordinary Intelligence (1992)
MASTERS JOHN	Bhowani Junction (1954)
MASTERS JOHN	Bugles and a Tiger (1956)
MOON SIR PENDERAL	The British Conquest and Dominian of India (1990)
PLANTERS ASSOCIATION OF CEYLON 1854 - 1954	(1954)
RUTHERFORDS PLANTERS NOTEBOOK	(1926 Ed)
SCOTT J. M.	The Tea Story (1964)
STANDISH ROBERT	Elephant Walk (1949)
SCHOFIELD VICTORIA	Every Rock - Every Hill (1984)
STEVENS G. R. LT. COL. OBE	Regimental History, The 9th Gurkha Rifles (Vol.2)
STILL JOHN	Jungle Tide (1930)
TUKER LT. GEN. SIR FRANCIS, KCIE, CB, DSO, OBE	Gorkha. The Story of the Gurkhas of Nepal (1957)
WEATHERSTONE J.	The Pioneers. Early British Tea and Coffee Planters (1986)
WILLIAMS HARRY	Ceylon, Pearl of the East (1950)
WILLIAMS J.H.	Elephant Bill
WOOLF LEONARD	The Village in the Jungle (1913)
WOODHOUSE L. G. O.	The Butterfly Fauna of Ceylon (1942)
VILLIERS SIR THOMAS	Some Pioneers of the Tea Industry (1951)

GLOSSARY TO PART 2

BABU	Office clerk
BACSHEESH	Tip, gratuity
BADMASH	Ruffian, bad character
BEDRAGGA	Tribal escort
BURQAH	Muslim woman's all-enveloping cloak with net eye-holes
CHAE	Tea
CHAGUL	Canvas or skin water container
CHATTY	Earthenware pot
CHOTA	Small
CHOTA HAZRI	Early tea (lit. little breakfast)
CHOWKIDAR	Night Watchman, caretaker
CUMMERBAND	Wide waist band
CHIT	Note, letter sent by hand
DACOIT	Armed robber
DAK	Post
FAQIR	Religious mendicant
GASHT	A patrol, usually by N. W. Frontier Scouts
GHARI	Horse drawn cariage
GHAT	Mountain pass or range
HAVILDAR	Sergeant
JEMADAR	Junior Viceroy's Commissioned Officer promoted from the ranks
JHODPURS	A type of trouser
KAREZ	A system providing permanent running water in arid country
KHASSADAR	Irregular tribal levies on N. W. Frontier
KHUD	Steep mountainside
KUKRI	Gurkha curved knife
KOT	A building used by the Kotwali i.e Police Supt. or such
MAIDAN	Any open gound or public park, a parade ground
MALIK	A tribal headman
MATI	Earth, soil
MULLAH	Musulim preacher
MUNSHI	Teacher, interpreter
MURGI	Chicken
NAUTCH	Dance
NULLAH	Dry river bed, ravine
PAGRI	Cloth used to make up turban
PANI	Water
PARATA	Type of unleavened bread
POSTEEN	Sheepskin coat with fleece on the inside
POWINDAH	Nomads
SHIKAR	Any field sport, especially shooting and hunting
SUBEDAR	Senior Viceroys's commissioned officer promoted from the ranks
TANGI	A gorge with very steep sides
TONGA	Two wheeled horse drawn trap
ZAMINDAR	Land owner

GLOSSARY TO PARTS 1 AND 3

ALAVANGOE	Crowbar
APPU	Butler, cook
BURGHER	Person of Dutch descent
CONDUCTOR	Senior Member of field staff, oftern in charge of a division
CUMBLY	Blanket, usally of thick natural wool
DURAI	Master
DURAISARNE	Lady
GHEKKO	House lizard
KACHCHERI	Office of a provincial administrator (Govt. Agent)
KADJAN	Frond of coconut palm
KANAKAPILLAI	Senoir field overseer lit. Accountant
KANGANY	Overseer, foreman
KATTY	Knife
KUTTI	Girl
MAIDAN	Any open ground or public park
MAMOTY	Inverted spade
MUTHALALI	Leader, top man
PANDAL	Archway, canopy
PERIYA	Big
PERUNARL	Holiday lit. big day
PESSASSEY	Devil
PODIAN	Boy
SINNA	Small
SOONDU	Tin, a tin used as a measure
SWAMI, SAMI	God
SYCE	Groom, horsekeeper
TARGRAM	Tin (metal) zinc roofing sheet
TAPPAL	Mail
TODDY	Juice of the palm flower, intoxicating when fermented
TOTAM	Estate, garden
TUNDU	Note, piece, bit
VETTA KATTY	Large knife, billhook

INDEX